FOR THE LOVE OF COCKROACHES

HUSBANDRY, BIOLOGY, AND HISTORY OF PET AND FEEDER BLATTODEA

Orin McMonigle

with contributions by

L. M. Roth

寵物與飼料蜚蠊之飼養、生物學、自然史教本

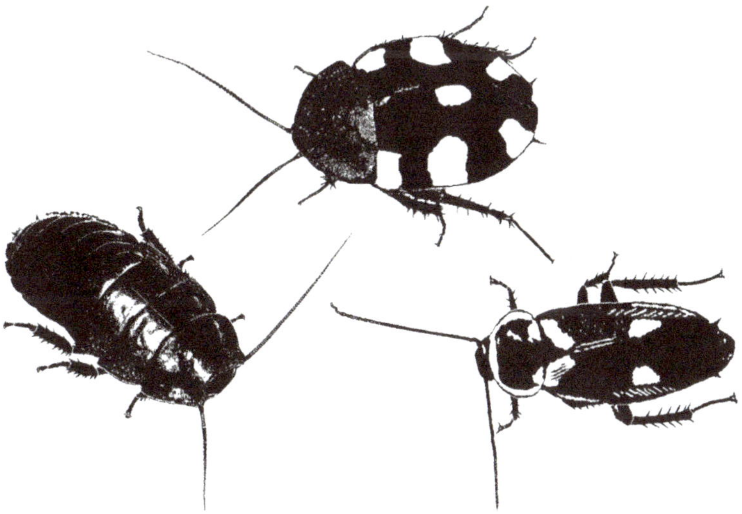

COACHWHIP PUBLICATIONS

GREENVILLE, OHIO

DEDICATION

This publication is dedicated to the late Dr. Louis M. Roth who provided text for the cockroach phylogeny and reproduction section, shared details on a variety of species and genera, mailed me copies of dozens of papers, and offered editorial suggestions for the original Allpet Roaches handbook. His contributions to understanding the biology of cockroaches and their taxonomy are legend. Dr. Roth was a tireless, if lesser known, promoter of keeping pet cockroaches. Every keeper who maintains *Schultesia lampyridiformis* and *Simandoa conserfariam* owe him not just for the descriptions of the animals, but also their collection. He established the *Blattella germanica* 'prowing' laboratory stock and gave hissing roaches away as pets for decades. A decent percentage of extant stocks may have originated from cages in his office.

For the Love of Cockroaches

ISBN 1-61646-427-5
ISBN-13 978-1-61646-427-1

CoachwhipBooks.com

Front cover: *Blaberus giganteus*. Back cover: *Corydidarum pygmaea* (left), *Therea olegrandjeani* (right).

CONTENTS

ACKNOWLEDGMENTS

I extend my greatest appreciation to Christian Schwarz, Gillian Higgins, and Peter Clausen for comments and suggestions on the structure and content of the text. Special thanks goes to Harley Rose, Glenn Holbrook, Ingo Fritzsche, George Beccaloni, and Edward Baker for providing personal experiences or resources for individual species accounts. Thanks also goes to Tom Larsen, Chen Zha, Pernille Aalvik, Tanaka Hideki, and Tibor Ludányi for providing some details. I appreciate the support from my daughters Kree and Gwynevere and my wife Sylvia. Lastly, thanks to God for my passion for cockroaches and a fair occupation that allows me to spend my free time learning and writing about these fantastic animals.

Cockroaches have been around for more than 360 million years. Over 4,600 species have evolved into various shapes and fantastic colors. While many roaches have adapted to live on flowers, inside rotten logs, in deserts and a multitude of other environments, only a handful are able to survive in our homes (without help). Many of the non-pest species make wonderful pets or even great pet food. This text is hoped to advance the knowledge and love of cockroaches.

蟑螂已存在超過三億六千萬年，並已進化出超過四千六百種不同形色物種。雖然有無數蟑螂生存在花叢、腐木、沙漠及多款生態環境中，然而只有少數能夠自立存活在我們的家居環境。許多非屬害蟲的物種可以當成極好的寵物甚至是絕佳的寵物飼料。希望此教本能夠提供對於蟑螂更進一步的知識與喜愛。

Eurycotis opaca “Giant Yellow-Margined” from Cuba

FOREWORD

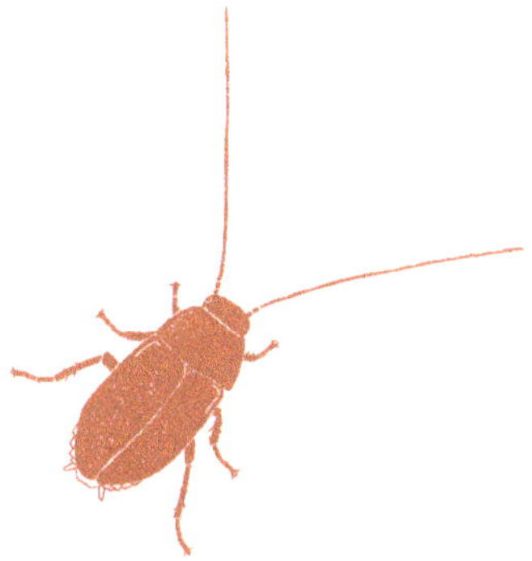

It was a very different world when Orin and I first became acquainted. It was 1992 and I was 14 years old. Back then, there was no Internet and even insect breeders from the same country did not know of each other's existence, let alone get in touch with foreign enthusiasts. There were no books on insect breeding, either. Whatever questions insect breeders had, they were on their own to figure out the answers. Given this background, insect breeders of the time were trained to be resourceful and critical thinkers. Without a doubt, Orin is a truly iconic insect breeder. He is the first person to have raised Goliath beetle larvae to adulthood. Today, anyone in the world raising Goliath beetles uses protein pellets, such as dog food, a technique developed by Orin.

In the summer of 1993, I visited Taiwan, where I was born, for two months. Orin called me almost every day from the United States to see what beetles I was catching. At the time, telephone companies had a monopoly on communication because there was no email, and international calls to Taiwan were $1 USD per minute. Orin spent $500 USD on phone bills that summer. This shows how passionate Orin is about insects.

Orin is a multi-talented insect breeder. Whereas my specialty is limited to beetles, Orin specializes in all insects, even encompassing other invertebrates including scorpions, spiders, centipedes, and millipedes. I sincerely hope Orin continues to work with all invertebrates and share his passion with the world through his well-written books.

In 1993, he told me that he had some pet cockroaches and offered to send me some. I was shocked and terrified! In my limited knowledge then, I thought that all cockroaches were hardy, dirty, stinky, and ugly. "Why would anybody want to keep them?" I thought. But Orin quickly explained to me that the vast majority of cockroach species do not stink and can only survive in pristine environments; in fact, some are colorful and pretty. I was doubtful. Again, do not forget Google did not exist. Nothing could be verified until I saw it with my own eyes. Orin started off by sending me some Madagascan hissing cockroaches. I vividly remember the first thing I did was take a whiff to see if they stank. They did not! I was also very impressed by their size and horns. Their live births caught me off guard, too. After I had learned the basics of roach husbandry, Orin followed up with sending me more species. I was blown away when I opened the package! The vibrant, neon green banana roach, *Panchlora nivea*, totally defies any stereotype of cockroaches. As its name suggests, the death's head cockroach, *Blaberus craniifer*, carries patterns on the thoracic plate that are reminiscent of the image of Satan. The translucent giant cave roach, *Blaberus giganteus*, is one of the largest cockroaches in the world, with a wingspan of 5-6 inches. The above were only some of the roaches Orin sent me. The variety and beauty of these cockroaches was truly eye opening.

Orin published his first book, *The Complete Guide to Rearing Grant's Rhinoceros Beetle*, in 1998 when I was a college freshman. I was jumping for joy when I received it in the university's post office. Passing students must have thought that I was going crazy. In fact, Orin's first book inspired me to write my book *For the Love of Rhinoceros and Stag Beetles.* Time flies. It has been more than two decades since Orin first sent me some Madagascan hissers. In this time, Orin has never ceased working with cockroaches. He published *Allpet Roaches: Care and Identification Handbook for the Pet and Feeder Cockroaches*, a complete breeding manual covering more than sixty species of cockroaches, as well as shorter publications on Blattodea. I am ecstatic to hear that Orin will be publishing this new book on cockroach breeding. Whether you are a beginner or an expert, with Orin's experience and meticulousness, I know this new book will be most invaluable to anyone interested in roaches!

Jonathan Ting Lai
March 15, 2016

The white eye mutation of *Periplaneta americana* is probably the oldest surviving captive blattodean bloodline.

A blue-margined roach from Singapore © Melvyn Yeo

Ellipsidion humerale from Western Australia © Jean and Fred Hort

Banded cockroach, *Cosmozosteria bicolor*, from Queensland, Australia © Bruce Thomson

Polyzosteria cuprea from Western Australia © Jean and Fred Hort

INTRODUCTION 介紹

The goal of Allpet Roaches (website, handbook, forums) has been to advance our hobby and to inform the countless cockroach enthusiasts that they are not alone. The original *Allpet Roaches: Care and Identification Handbook to the Pet and Feeder Cockroaches* (McMonigle and Willis 2000) was written to promote the worldwide keeping of roaches and to provide a quick reference guide for the cockroach enthusiast. It has been cited in various texts, has sold nearly two thousand copies, was translated into Italian, and served its purpose. However, the handbook was small and the format greatly limited its content and distribution. The pre-ISBN version stated the ultimate goal was to produce a large-format, color version, but this statement was removed from the ISBN version (2002) since it seemed to be pure folly. This book, *For the Love of Cockroaches: Husbandry, Biology, and History of Pet and Feeder Blattodea*, is the large-format, detailed, full-color hardcover which was the original but impossible goal for the Allpet Roaches handbook.

Petting Zoo. This was part of a big bug festival (3,000-10,000 attendance) that began in 1991. The petting zoo was a popular exhibit that ran from 1994 to 2014, the last year of the festival. I usually used trays with *Blaberus craniifer* and *Archimandrita* since these species cannot climb the sides.

What took so long? This text has taken more than sixteen years since the original Allpet Roaches handbook was written. For more than a decade I did not think a massive, color hardcover on cockroaches could ever be possible without personal bankruptcy. After it became a possibility with Coachwhip Publications just a few years ago, it still seemed like an impossible amount of work for something that would barely make a whisper in terms of copies in circulation. Moreover, my first large-format text, *Invertebrates for Exhibition* (2011), already offered this order of insects reasonable coverage. One of the longest chapters and the cover photo were decidedly Blattodean. Assembling a full text on this group of insects is a daunting, almost frightening task for me because I want to do a good job and it is difficult to effectively cover cockroaches in the way

Previous publications

these fantastic animals deserve. It is difficult to distill thirty-six years of experiences with a hundred species without making the text too long. Compared to most of the other less popular invertebrates, there is a far more massive backlog of associated literature to compile and sift through. Incredible volumes of information were written before the turn of the millennium. Most were impossible to acquire, even with access to a university library, when the Allpet Roaches handbook was written and nothing more could be squeezed into that format anyway. The sheer volume of material I have read since and would attempt to incorporate is overwhelming. Then, there are the inconsistencies in phylogeny, systematics, and taxonomy, unparalleled in many invertebrate groups. This book emphasizes husbandry explanations, culture data, and personal experiences which are immutable and cannot be changed at the whim of the next worker.

When it comes to cockroach books, there have been a few popular joke texts like *The Compleat Cockroach* (Gordon 1996) and *The Cockroach Papers* (Schweid 1999), but these limit discussion to a few scientific facts, pages of silly roach anecdotes, and a wholesale appeal to irrational fears. A detailed text on cockroach biology was published in 1960 (by Roth and Willis), with the concept expanded and updated in 2007 (Roth and Willis 1960; Bell, et al. 2007). Bell also published a much shorter book on the 'laboratory cockroach' (Bell 1981). These black and white textbooks are devoted mostly to interesting biological anecdotes and theories, studies on pest roaches, and species that are not maintained in captivity. Certainly all the available books are informative and may be fun to read, but they do not provide information for the multitude of people who keep roaches as pets or as feeder insects. Starting in the early 1990s, a variety of general invertebrate texts included a few paragraphs to a few pages of very limited information on cockroach care. In the years since the original *Allpet Roaches* handbook was written, there have been texts with longer and better sections than before, but the only moderately extensive cockroach treatment (in English) is a chapter in *Invertebrates for Exhibition* (McMonigle 2011e). An 80-page, A5 format, German-language, full color text on roach husbandry was published a few years ago (Bruel & Nake 2011). Rentz (2014) authored a handsome full-color, soft cover guide to identify Australian species with a short basic husbandry chapter written by Deanna Henderson. To date, there has never been an English-language, large format, full color text devoted to the Blattodea, and certainly not one with the enthusiast in mind. *For the Love of Cockroaches: Husbandry, Biology, and History of Pet and Feeder Blattodea* fills this gap in the cockroach literature. The intent is to provide a detailed biology text, reproductive husbandry manual, and color picture book in a combined format.

Although there has been extensive information available on pest cockroaches, there has been limited information available to the dedicated enthusiast who is interested in non-pest species. There are thousands of scientific papers that discuss research on neurology, hormones, locomotion, resistance to environmental variables, pesticide effects, memory, and genetics. Myriad published accounts detail how long it takes these creatures to die after being poisoned, smashed, or decapitated and deprived of water, air, and food. Published research papers usually only provide information relative to the most widespread of the pest species (the American cockroach *Periplaneta americana*, the brown-banded cockroach *Supella longipalpa*, the Oriental roach *Blatta*

Therea olegrandjeani, a beautiful cockroach

Although there are incredible numbers of small, discreet roaches from warm temperate areas to the tropics, this text concentrates on large, common pets. *Cariblatta lutea* female from Montgomery, Alabama, found scurrying on the forest floor. © Alan Jeon

Cockroach from Ecuador (CC-BY) Graham Wise

orientalis, and the German cockroach *Blattella germanica*). Papers relating to species descriptions, taxonomic hierarchy, and generic redesignations are legion and cover every known species. Of course most taxonomic literature is restricted to measurements of body parts and descriptions of dissected genitalia. Scientific papers concerning laboratory culture or natural history for species appealing to hobbyists, especially for desired genera like *Megaloblatta* and *Paratropes,* are extremely difficult to acquire (if such papers exist).

BLATTODEA CULTURE GROUP

Since the turn of the century, articles in hobbyist magazines and invertebrate journals have covered the majority of cultured species. *Invertebrates-Magazine* (United States) may be the most prolific with three dozen articles, but there have been a number of articles in *Sklipkan* (Czech Republic), *Exotiske Insekter* (Denmark), and *Arthropoda* (Germany). Furthermore, there are more than a dozen species accounts in the *Blattodea Culture Group* (United Kingdom) publications dating back to 1986 and the more recent *American Cockroach Society* publications. It would take a lot of time and money to track down all the periodical literature published over the last few decades. This book attempts to compile all available husbandry data along with unique experiences from the author. Pest species and tiny native species are discussed but they are not meant to take center stage.

This compendium is about the love of cockroaches so those species kept as pets and display animals are featured. The title refers to the great fascination and appreciation hobbyists have, and others should have, for cockroaches. It is also a nod to a previous text credited with bringing new life to the culture of Coleoptera. The title is not a nod to *For Love of Insects* (Eisner 2003), but rather *For the Love of Rhinoceros and Stag Beetles* first edition (2001) written by Jonathan Lai and myself.

American Cockroach Society

SPECIMEN IDENTIFICATION AND COCKROACH TERMINOLOGY

Nearly all of the commonly kept cockroach species are easy to identify with a clear photograph of the adult (rarely with a nymph or ootheca photo). Those species whose identification is more complex are covered in the species list (specifically *Blaberus* and *Gromphadorhina*). Anyone who keeps cockroaches is likely to learn at least a little about their taxonomy and identification. Here we will try to get the beginner familiarized. The original seven-tier system and binomial nomenclature

Tropical polyphagids are popular pets today because they are colorful, but even the majority of tropical species are brown. This unidentified female polyphagid burrows in the forest floor of the rainforest in Panay, Philippines. © Christian Schwarz

This is the same species from the Philippines, male. © Christian Schwarz

has changed to encompass dozens of levels and more complex names. Subspecies are common to some other arthropod taxonomic schemes, but are not usually popular with blattodean taxonomists. This text will only delve lightly into the standard tiers. Massive databases could be filled with taxonomic changes, rules, and information. This book is intended as a guide for the enthusiast and includes taxonomic information commonly encountered by those interested in studying, observing, recording, and enjoying the development, lifespan, and reproduction of living cockroaches.

Classification affects even the casual keeper. Sometimes part of the scientific name or its translation is used as the common name. Some, especially new stocks, are referred to by their scientific names since only common species can have common names. Taxonomic identifications are not always perfectly defined due to variation between specimens or populations; others are undescribed. Some species have different morphological characteristics (beyond color, which means nothing when inconsistent but is used in many descriptions) than closely related species. Closely related species sometimes can be made to interbreed with little prompting in laboratory settings and continue on for generation after generation. This may or may not call into question the validity of species identification or description. Enthusiasts, at the least, want to know what something is to be called and what must be kept separate.

Uncertain identifications (guesses) should be placed in "..." or end with a question mark. In the last few years it has been popular for vendor hobbyists to pull a scientific name out of a hat for an unidentified stock and add cf. to the label. The cf. qualifier is short for *confer* or *compare to*. However, unidentified species are simply given the names of real species that may or may not look similar or may not be related. The cf. qualifier is not supposed to be used for uncertain or doubtful identification but rather for uncertainty in the taxonomy. It is supposed to mean "with some certainty" rather than "not a clue."

Even the common cockroach can be a rather cute and curious creature in the right light: *Paratemnopteryx couloniana*.

Names and the year of description are included within the species lists to give context to the names since the taxonomist and year can carry interesting details in themselves. These are additional to the binomial scientific name and the extra details they provide cannot change the validity (or lack) of a species identification. As standard, if the genus was changed after the original description, the name of the author is in parentheses.

Terminology in this text will follow commonly used words as much as possible. The word parturition will not be used instead of

Mitchell`s diurnal cockroach, *Polyzosteria mitchelli*, from Western Australia © Jean and Fred Hort

birth, nor apterous to replace wingless, nor haemolymph to avoid blood. However, there are no common words for terms like cerci, arolium, and sternite. Also, some ideas cannot be translated into a few simple words. For example: ovoviviparity and viviparity are different means of reproduction, but both could translate to live bearing, while egg laying and egg carrying are both oviparity. A glossary is provided at the end to bring the beginning hobbyist up to speed.

THE INTENT OF ALLPET ROACHES

In the late 1990s it seemed more and more invertebrate books would contain expanded sections on cockroaches, but there has been very little progress. Unfortunately, the only English in-print book devoted to the care of cockroaches is the *Allpet Roaches Handbook* (2000, 2002). It is limited in scope by its construction and was written more than fifteen years ago. This text is an exponentially expanded and updated version of the handbook. These pages contain information on history, biology, diet, reproduction, captive habitats, list of available species, and aspects related to feeding roaches to other pets. Information is provided in a straightforward manner so that amateurs and experts alike can better understand and enjoy these creatures. The large number of cockroach enthusiasts should have a book available to them that answers husbandry questions and displays a "wish list" of incredible and beautiful species. Those hobbyists who presently only keep one or two species of roaches may be inspired to hunt down some of the incredible species depicted and discussed within.

Paratropes sp. in rain from Venezuela © Arthur Anker

INFORMATION BASIS

When reading a book, especially one with husbandry suggestions, one must ask him or herself if this information is going to be helpful or harmful. What is the source? Of course if the information is helpful it will be thousands of times more valuable than the cost of the book and expense of the time spent reading it. A single valuable tip could save an irreplaceable animal or culture stock. If the suggestions are just unproven hearsay they could prove deadly expensive. The author of this book has reared generations of difficult cockroaches like *Macropanesthia, Corydidarum*, and *Gyna capucina*, and has maintained consecutive cultures of dozens of species over decades. Maybe a handful of people in the world could boast equivalent depth and breadth of husbandry experience. He has published dozens of articles on husbandry for different species and composed *Allpet Roaches: Care and Identification Handbook to the Pet and Feeder Cockroaches* (2000). The author was even the American Cockroach Society president at one point in time. For a time he seemed to be a very important man in the blattodean world, but then one day somebody said, "Hey, you're just a big cockroach."

Blattodea fossil (Museo Civico di Storia Naturale di Milano)
(CC-BY) Paleo Bear

MAN AND COCKROACH 人與蟑螂

MAN AND THE FAMOUS COCKROACH

If one can tell from creation that God has an inordinate fondness for beetles as a result of their variety, one might also conclude cockroaches were the original favorite. Many may look like beetles, but they were first, so beetles generally look like them. Roaches are one of the few animals to look similar to fossils after 360 million years and are found across the majority of habitats globally. Nevertheless, there is some debate over which fossils were roaches or simply roach-like insects because a number of the oldest fossils had long ovipositors like a cricket.

This fossil cockroach from the Crato Formation in Brazil shows the underside and one tegmen of what could be mistaken for a modern cockroach. This individual however lived around 115 million years ago, close to a tropical lagoon. © Edward Baker

The word *cockroach* is a transliteration of the Spanish word *cucaracha* made popular by a satirical folk song. One version of the Spanish folk song *La cucaracha* speaks of a cockroach who cannot walk because he ran out of marijuana. The song is thought to reference various political leaders, but who is uncertain because the song may date back 100 or 500 years. The Spanish word *cucaracha* is at least as old as the song might be and may come from the term *cuca* for caterpillar. The shortened form of cockroach (roach) is used more commonly but has a few other meanings. It is common slang for the butt of a marijuana cigarette. A roach is also the name for a hardy food fish (*Rutilus rutilus*) common in Europe. In most literature the word roach and cockroach are used interchangeably.

The Order Blattodea is the most hated of the insect orders in Western cultures, even beating out lice, flies, and fleas. Although blattodeans should be considered among the least worrisome of pests, pest control companies successfully employ fear of cockroaches above any other pest to sell their services. Many Eastern and African cultures have not found roaches to be bothersome and a very few cultures enjoy them as food. (Though healthy for the human digestive system, many species have a rather unpleasant aftertaste or spiky legs.)

Everyone in Western society is associated with the stereotypical fear of roaches. The stigma of roaches living among personal possessions bothers even Westerners who possess no irrational fear of the animals themselves.

The general public's deeply rooted ignorance of cockroach biological diversity and beauty can be disheartening for the cockroach enthusiast. At the same time, there seems to be a respect associated with the great and admirable staying power of the roaches. Everyone knows they were around for the rise and fall of the dinosaurs and expects them to exist after humans are wiped from the earth. This knowledge has taken on a life of its own in stories about roaches living without heads, surviving nuclear bombs, and reproducing without food or water.

A very common urban legend or old wives' tale that promotes fear of pest roaches is, "cockroaches spread disease." This has not been demonstrated. Some experiments have shown the outside of the exoskeleton can be coated with bacteria that survive to subsequent testing or specimens from a contaminated environment can test positive for external bacteria, but never a proven case of transmission to humans (Roth and Willis 1957a; Cochran et al. 1975; Burgess 1979; Carpenter 1992). The idea always seems to be that transmission may be possible despite lack of direct evidence (Fotedar et al. 1991). One may prove a blue whale will crush a human to death if it falls on one, but using such reasoning to promote the extermination of blue whales due to the associated health risk is no more unreasonable than the bad rap roaches receive. By design, correlation of health effects with pesticide use versus the cockroach is an afterthought if considered at all. By the mid-1900s a slew of scientific papers inaccurately blamed cockroaches for the transmission of parasites or amoebas causing malaria, hepatitis, various types of cancer, and even disease caused by vitamin deficiency including scurvy (Roth and Willis 1957a). Other pests, specifically rats and mice, carry vertebrate diseases and have been shown to transmit hundreds of illnesses to humans because they are also warm-blooded vertebrates. Pet cats are common vectors of bacteria or protozoan diseases such as toxoplasmosis. Mammals of any size are infected by many of the same or related disease organisms. Blattodeans have a very different physiology from humans and are not infected by vertebrate diseases. While there are arthropods that feed on vertebrate blood like tsetse flies, mosquitoes, fleas, ticks, and bloodsucking true bugs which transmit vertebrate diseases through vertebrate body fluids, cockroaches do not feed on vertebrate blood. The idea that roaches could help to transmit disease seems to be rooted in the unlikely possibility that they might walk across a contaminated surface in a structure and then later, across food. Cockroaches are not filth flies that land on people, food, and feces. The transmission of pathogens is caused by physical contact with an infected person, animal or their bodily fluids, not by the minuscule surface area of blattodean tarsi hiding beneath the furniture. Furthermore, when pest species consume pathogens, even certain bacteria by the billions, many types are entirely digested by the cockroach (Roth and Willis 1957a). Unfortunately, ignorant misconception, agenda-based science, and propaganda cause cockroaches to be targeted when the efforts should be spent on credible disease vectors.

Cockroach-induced asthma is a pervasive story repeated by pest control companies to sell their chemical wares. The myth is repeated by ignorant health professionals and researchers who have not read the full text of the studies. It is true some studies have shown an increase in the severity of asthma attacks in persons who are allergic to cockroaches and already have asthma. (A percentage of the population is subject to cockroach sensitization, but cockroach allergies and asthma are not the same thing.) Cockroaches do not cause asthma and asthmatic patients who do not have cockroach allergies are not affected by high levels of cockroach allergens (Rosenstreitch et al. 1997). Of course linkage of asthma to pesticide application is little researched, if not wholly ignored in most studies. Even where research considers cockroach allergies a factor, sensitivity to outdoor allergens is more prevalent (Sastre et al. 1996). A massive study on immunology recently completed showed that exposure to cockroach allergens early in life actually protects babies from asthma (Lynch et al. 2014).

Blaberus giganteus at the Audubon Insectarium in Louisiana

The reproduction myth is an interesting urban legend based on a mathematical function. This story suggests an incomprehensible number of individuals are the practical result of a relatively few generations. Sadly, even some insect zoos promote this story (Orkin Insect Zoo exhibit, National Museum of Natural History). The idea of exponential growth is where in a relatively small number of generations the final total weight of theoretical offspring would eclipse the mass of planet Earth. Some cockroaches can reproduce in relatively large numbers, but they are limited by many factors, usually lack of resources as much as predators. Elephants, humans, and blue whales could also grow to eclipse the mass of the entire earth after a few dozen generations were there no limiting factors.

Among the most common stories is that cockroaches 'live' for weeks after their heads are cut off. Of course this story is only somewhat true with a rather liberal interpretation of the word 'live.' An animal without a head, whether it be a chicken running around the barnyard, a recently guillotined human, or a cockroach, is not alive by most definitions—it is dying. Cockroaches have open circulatory systems like all arthropods and, like most arthropods, severed parts of the body occasionally survive for days or weeks without circulation. However, millipeds are the champs in this area (120 days for an *Archispirostreptus* without a functional head or mouth, McMonigle 2012a). Still, fading slowly is an exception rather than a rule. If a cockroach's head is accidentally smashed in the cage lid, the body commonly stops moving within seconds.

The idea that cockroaches will survive a nuclear holocaust likewise takes certain liberties. First, there are no cockroaches capable of surviving the intense heat or shockwave forces of a nearby explosion. A global nuclear winter would kill off around 95% of cockroach species, since the majority are tropical. The temperature alone would kill them even if their habitats and food were not utterly decimated. In areas less ravaged by the immediate effects, radiation would still be a stumbling block. Adult cockroaches are capable of surviving high levels of radiation because they are not growing, but nymphs and eggs would not develop. A few species may develop at levels of radiation deadly to most other organisms. However, roaches are not the only organisms capable of this. Deep-sea life would probably be the least affected by an all-out nuclear Armageddon.

One myth with hardly a hint of truth is that the roach and its oothecae are nearly impossible to smash. The exoskeleton is pliable so there is some give, but not enough to save any animal from a half-hearted attempt. The main pest species are especially frail and gentle handling of small nymphs often results in their damage or immediate demise. It is very easy to smash an ootheca and even if the damage appears minor, none of the eggs will survive. A tiny crack dooms an ootheca, let alone the compression and liquidation of the eggs. This story may come from people confusing cockroaches with fleas, dead and live egg cases appearing similar, or the propensity for man to spin a good yarn.

The giant cockroach myth is a whimsical story and a fabrication that may date back more than 300 million years. I have displayed *Blaberus giganteus* (~84 mm) at various events and have heard countless times how the palmetto bugs (almost never more than 40 mm in length) seen down south, or smaller species in northern basements and sewers, seen years earlier were bigger. Human inability to accurately determine and remember size is the reason standardized measuring devices were invented. In 2004, the discovery of a new species in Borneo was widely reported and the titles of various articles claimed it was the largest cockroach in the world (at 4″ or 100 mm body length). However, the few specimens of this species (*Miroblatta baai* Grandcolas and Deharveng, 2007) thus far measured are 60 mm long.

"The cockroaches have been with us for a long time: huge cockroaches crept through the dank vegetation of the ancient world and, little changed, their smaller descendents have followed us through the slow course of history down to the present."

—Hutchings 1966

The giant cockroach myth is rarely based on something as concrete as this photo. The quarter is a fake miniature coin replica, while the roach is a very large *Gromphadorhina* sp.

The myth of prehistoric giant cockroaches is an idea with less certain beginnings that showed up in entomology texts long before a moderately large fossil was ever discovered. Most available fossils are tiny (10 mm or less). The largest fossil cockroach measured by man, *Xenoblatta scudderi* from a coal mine in eastern Ohio (Easterday 2001) is 79 mm, plus an additional 10 mm for hind wings that seem to have been unnaturally splayed out past the tegmina (5 mm shorter than the normal pronotum to tegmina length of a female *Blaberus giganteus*). At least one living species of *Megaloblatta* from South America measures 100 mm to the end of the tegmina (Gurney 1959b). A

Without the wingspan, *Blaberus giganteus* are still large cockroaches.

Size quotes are almost never backed by measuring equipment. This giant stock hisser is probably the largest documented hissing cockroach (2015).

These male hissers have hyper-extended abdomens from a large meal. Usually this is seen only as a result of bacterial infections that cause bloating or shortly after the rare overfeeding event (it is difficult to get them to eat this much). Although the upper male is much smaller, he measures almost the same as the large male because of a more extended abdomen.

Measurement and the hyper-expanded abdomen: this condition is caused by an infection and adds about 10 mm to the natural length. This animal can still live a long life if provided adequate ventilation.

Blaberus giganteus is the largest winged cockroach in captivity. Even a small female like the one pictured measures nearly half a foot across.

large *Macropanesthia* or exceptionally large *Gromphadorhina* can measure in at 80 mm body length, but they are wingless so the telescopic nature of the abdomen means a fossil (or over-fed specimen) can be made, with minimal cheating, to measure a centimeter or two longer.

Man has introduced cockroaches throughout the world. Since many introductions happened hundreds of years ago, the origin of most adventive species will never be known with certainty. Even the most recent introductions have uncertain beginnings. The tiny *Luridiblatta trivittata* (Serville, 1838) found in northern Africa, southern Europe, and the Middle East, showed up in California, USA, about ten years ago. Introduction of alien species is certainly undesirable but it will continue as long as man ships food, plants, and products around the globe. Accidental introductions are not a major concern because cockroaches are at the bottom of the food chain and so are unlikely to damage food webs or eliminate other species, but also because there is not great value in worrying about the inevitable. Common adventives are associated with disturbed areas and are found with man. They are almost never found in undisturbed habitats. Most or all of the adventive cockroaches would disappear if man disappeared.

ROACHES IN POP CULTURE

Cockroach references in movies, television, and video games could take up hundreds of pages. The rarest movie would be the apocalyptic flick that does *not* show or reference the cockroach. The most deservedly famous cockroach movie is *Joe's Apartment* (1996) starring life-like talking *Periplaneta*. There is also a surprisingly long list of books, paintings, and sculptures based on cockroaches.

While dozens of species have been portrayed in film, the Madagascan hissing cockroach is

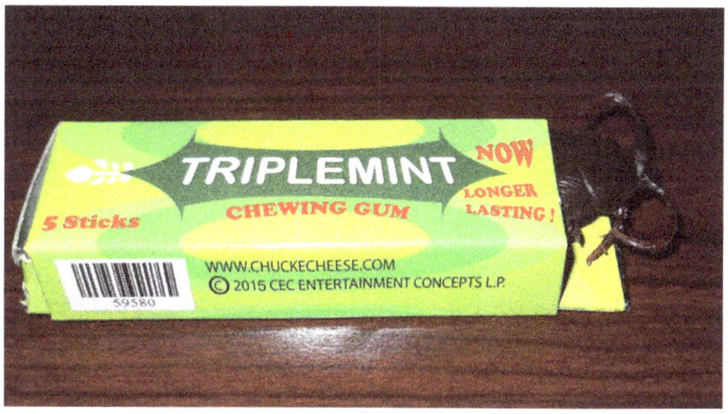

A gag gift: man`s fascination with fake cockroaches is endless.

the undisputed king of film and television. The first time I saw the hissing cockroach in a movie was *Bugs* (1975), when they crawled out of a crack in the earth following an earthquake. *Starship Troopers* (1997) showed hissers in a propaganda commercial. *12 Monkeys* (1995) and *Men In Black* (1997) also featured hissers.

From a 2013 Associated Press article about roaches as pets: Gwynevere with her pet cockroach, Dr. Phil.

The most widespread attention came from its appearance in dozens of episodes of *Fear Factor* (2001-2006, 2011, 2012). Hissing cockroaches were adorned with jewels and attached to leashes held by models in America's Next Top Model (2006).

HOBBY HISTORY

Rearing cockroaches may be among the oldest passions, certainly within the invertebrate realm. The oldest currently living stock dates back to specimens collected in a deep coal mine in 1944. These are the *Periplaneta americana* 'white eye' form. Before that, *Panchlora nivea* was maintained in captivity (Davis 1930), as was *Blaberus fusca* (Saupe 1928), but those bloodlines are almost certainly long gone. *Blaberus discoidalis* culture stock collected from Jamaica circa 1950 (John and Lewis 1958) may be the late 1970s stock still maintained today. Articles in the 1960s suggested *Blaberus* species make great pets and *Blaberus giganteus* have long been established pets kept in terraria (Banks 1969, Stanek 1969). Also some time in the 1950s, *Gromphadorhina* arrived in U.S. culture through Europe (Roth and Willis 1960). My first hissing cockroaches came from Dan Marino (Director of the Cleveland Aquarium) in 1980. He had maintained a colony of *Gromphadorhina portentosa* since 1972. He also had acquired *Blaberus giganteus* but his colony never did well enough to divide it. It was to be another fifteen years before enthusiasts would have the opportunity to acquire this massive, winged species with consistency. In the early 1980s, hissing cockroaches could be found at pet stores such as Parma Pets in Ohio and *Blaberus discoidalis* became popularly traded as reptile food. *Blaberus discoidalis* could be ordered from Carolina Bio, North Carolina, for years prior to 1982, but not at feeder prices. The 1980s was probably about the time Agama International's *Eublaberus* sp. 'Ivory' was established. The late 1980s and early 1990s brought increased popularity with species such as orange head 'beetles' offered by Special Care Pet Center in Pennsylvania. Purdue University giant strain hissers made their debut, but were likely a strain of the original 1950s stock. The green adults of *Panchlora nivea* became a popular feeder. The magnificent pet *Blaberus giganteus* were finally readily available through invertebrate vendors like Mascarino Tarantulas in California. By the late 1990s, death's heads and dozens of other species were common stocks maintained in the U.S., Europe, and Asia. Most of these culture stocks were not new captures, but rather had been maintained in laboratories and biology classrooms for years, if not decades, prior. As late as the mid-1990s the more zealous hobbyists could only amass ten different tropical species after ten to twenty years of collecting. Hobbyists had been few and far between before the turn of the millennia, but the internet made it possible to quickly locate others with similar interests. Today a hundred or so species are kept with some regularity and it would be difficult to maintain them all concurrently. Many species and many more bloodlines have disappeared from culture, but quite a few very old stocks still exist.

COCKROACHES DISPLAYED AT INSECT ZOOS

Blaberus giganteus can be seen at the majority of insectariums as it is arguably the single most impressive animal in culture. The giant hissing cockroaches are also seen in displays but are more popular as outreach because they are easily held by volunteers. Other species are not commonly seen at zoos, but Cincinnati Zoo's World of Insects still displays cultures of *Eurycotis decipiens* and *Panchlora nivea* dating back more than twenty years (Cincinnati Zoo Animal Inventory list 1993). The spectacular *Therea* species have made their way to insect zoos in recent years.

Unfortunately, even insect zoos whose responsibility is to educate the public sometimes pander to hype and hysteria and even manufacture fear. The 'cockroach kitchen' diorama

This giant cave cockroach display at the Cleveland Botanical Garden (Ohio) is part of a longstanding Costa Rican glass house exhibit.

Quite a few species like these tiny *Neoblattella detersa* (Walker 1868) from the West Indies have been transplanted by man with little fanfare or effect. These are adventive to Florida. © Alan Jeon

Sometimes pest roaches are seen in insect zoo displays as the zoos are usually located in large urban structures with limited pesticide usage. This appears to be a *Periplaneta americana* nymph in a *Blaberus giganteus* display.

The Madagascan hissing cockroach display at the Cleveland Botanical Garden (Ohio) is part of a longstanding Madagascan glass house exhibit.

at the Audubon Insectarium in New Orleans, Louisiana, is a cage filled with dollhouse furniture. It is arranged to look like a kitchen, complete with appliances, dishes, and silverware. The enclosure contains about 100 *Periplaneta australasiae* with their feces pasted all over the furniture and dishes. At the very least, full-size silverware and *Blattella germanica* should have been used to add a little authenticity to the display. The Victoria Bugzoo in Canada has dollhouse dioramas with *Gromphadorhina* running around. Similar displays are found in a few other insect zoos around the world and unexplained videos of 'giant' roaches taken by visitors show up now and then.

Open-top hissing cockroach display at Ark Avilon Zoo, Philippines (CC-SA) Keren Tan

Cockroach kitchen display at the Audubon Insectarium in Louisiana

USDA exhibit using Madagascar hissing cockroaches (USDA photo by Lance Cheung)

Cockroach display at Natural History Museum of Los Angeles (CC) Anika Malone

PLATE VI.

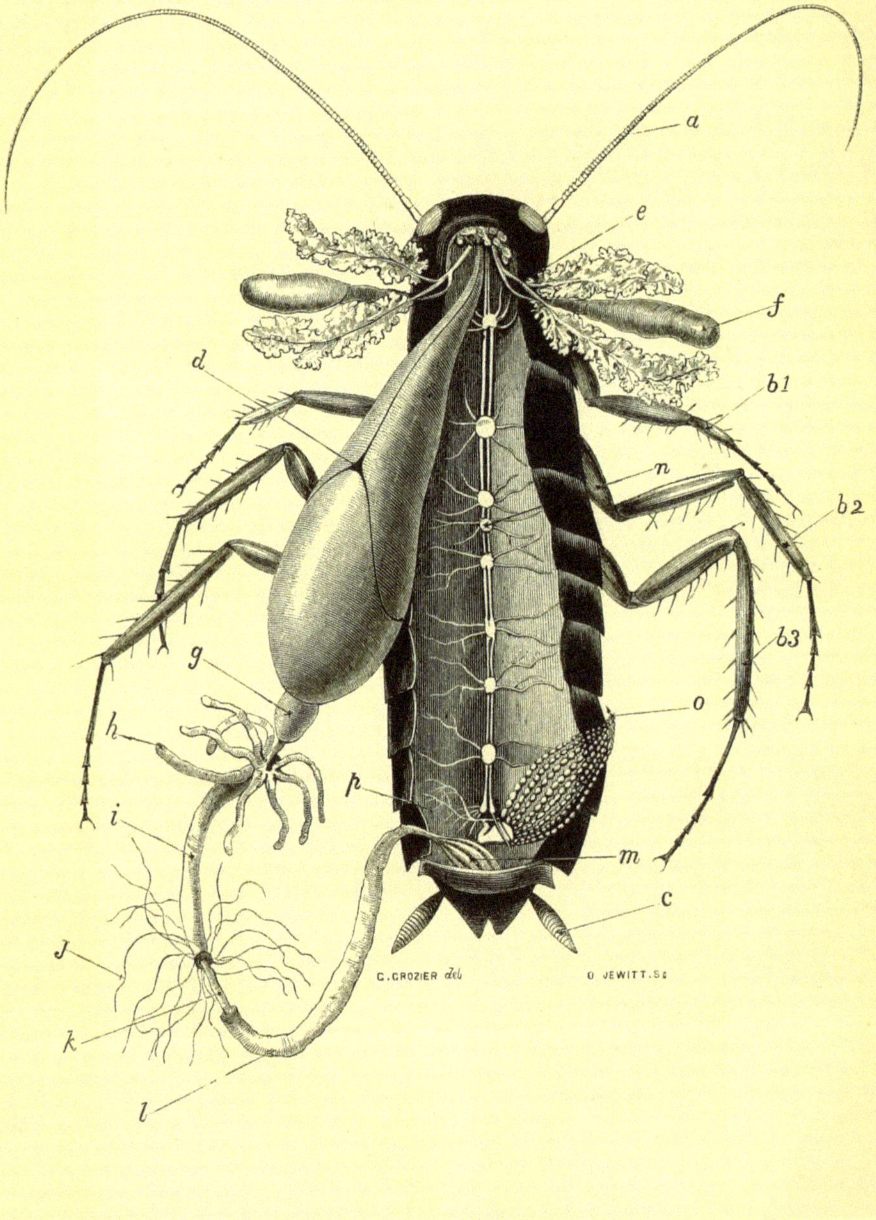

Cockroach, *Periplaneta Orientalis.*

Blatta orientalis, from Rolleston`s *Forms of Animal Life*, 1870

SYSTEMATICS, MORPHOLOGY, AND BIOLOGY

分類、型態、生物學

CLASS INSECTA

Superorder Dictyoptera

Order Blattodea
(cockroaches and termites)

Order Mantodea
(mantids)

The praying mantis, a beloved cockroach relative

Grasshoppers like this one, *Brachystola magna* from Arizona, are rarely associated with roaches by the general public, though they are closely related.

Cockroaches (Blattodea), praying mantids (Mantodea), walkingsticks and walkingleaves (Phasmida), grasshoppers, crickets, and katydids (Orthoptera), and the lesser-known rock crawlers (Grylloblattidae) had long been considered members of the same order, the Orthoptera. Currently the Blattodea and Mantodea are the only members of the superorder Dictyoptera. It is ironic that the most closely related order, the Mantodea, is not despised like roaches, but beloved. Few people make the connection between mantis and roach. Praying mantids rank among the most admired and loved of all insects.

On the other hand, it may be of no surprise to discover that cockroaches and termites have proven to be members of the same order (Inward et al. 2007). Members of the wood-eating roach genus *Cryptocercus* contain the same symbiotic flagellates in their hindgut as some termites and are more closely related to termites than to many other cockroaches. *Mastotermes* termites have the same foregut digestive fauna (genus *Blattobacteria*) as roaches and lay eggs in oothecae (Bell et al. 2007). Although termites have turned out to be eusocial cockroaches, they are traditionally treated separately from other cockroaches. Termites are not covered in this text because their husbandry is vastly different and they are seldom kept as pets. Worker termites of native species are commonly collected for feeding young dart frogs. *Zootermopsis* workers and soldiers have been offered by biological supply houses since the 1980s and rearing instructions for termites have long been available (Morgan 1989, 1999) but there is little interest.

Some animals are similar in shape to roaches, but are very distantly related, like this crustacean, a seashore isopod, *Ligia exotica*.

FAMILIES IN THE BLATTODEA

The number of families making up the Blattodea has been in a state of flux for more than half a century, from five to twenty-four depending on the author and day of the week. The relationships between them are greatly contested depending on who is doing the phylogenetic analysis (Grandcolas and D'Haese 2001). The Blaberidae, Blattidae, Cryptocercidae, and Polyphagidae have been widely accepted, as well as the Blatellidae, now being called Ectobiidae. Familial assemblages are academic but generally contain species with similar characteristics. This text is not meant to debate competing ideas of systematics or interpretations of phylogenetic analysis. The selected species chapter is in alphabetical order partly because family grouping may change again by the time this text is printed. Here we will consider the groupings according to Beccaloni's Cockroach Species File Online (cockroach.speciesfile.org, Version 5.0/5.0), with the exception of Corydiidae, where I have instead retained the longstanding Polyphagidae family as *Corydia* has been shown to be an invalid genus.

BLATTIDAE—The word *blatta* means cockroach in Latin and this family contains some of the most familiar large, egg-laying species. There are around 600 known species. Many are beautiful

Blattidae: *Eurycotis decipiens*

Blattidae: *Periplaneta americana* (USGS Bee Inventory and Monitoring Lab)

Blattidae: *Dorylaea orini*

Ectobiidae: *Chorisoneura texensis* female © Alan Jeon

Cryptocercidae: Large *Cryptocercus punctulatus* nymphs from West Virginia (2004) resemble termites.

forest and savannah dwellers, while a few are among the best-known structural pests. *Blatta, Deropeltis, Eurycotis, Neostylopyga,* and *Periplaneta* are well-known genera. Members of this family are mostly large-bodied species (greater than 1″ or > 25 mm), and a large proportion are wingless or short-winged. Family members are most readily identified by the female's pointed terminal sternite and the external valves used to form oothecae.

Ectobiidae—Until recently this was listed as the Blattellidae (*blattella* means little cockroach) and one of the most varied subfamilies is still called the Blattellinae. This family contains about half (2,300) of the 4,600 known cockroaches. Commonly kept species include those from the genera *Anallacta, Blattella, Ectobius, Parcoblatta, Pseudomops*, and *Symploce*. This family includes two of the most desired genera that have not yet been successfully kept, *Paratropes* and *Megaloblatta*. Though the majority of species are small at around half an inch (~12 mm) in length, some are large. *Megaloblatta longipennis* has the largest wing area of any species yet seen in the history of the earth. A small proportion of species are short-winged (for example *Parcoblatta* females) or wingless, but it is a large and variable family. Females form oothecae but the terminal sternite is rounded and there are no external valves. Some species carry the ootheca partly retracted in the abdomen until shortly before hatching and a few rare species are live bearing.

Blaberidae—Most of the cultured species, and arguably the most amazing species, are part of this family, also known as the live-bearing cockroach family. With about 1,200 species, this is the second largest family. Oothecae are only seen briefly because they are retracted into the body shortly after formation, an important but not easily seen identifying characteristic. The terminal appendages (cerci) are usually smaller in relation to the body compared to the previous two families and are often hidden by the wings when present. Genera include *Blaberus, Corydidarum, Gromphadorhina, Macropanesthia, Nauphoeta, Panchlora*, and *Simandoa*.

Dried specimen collection, mostly Blaberidae (one *Cryptocercus*)

Polyphagidae—These roaches are among the most beloved today, but were rarely seen in culture before 1995. This family has long been called desert or sand roaches because many species are perfectly adapted to desert life. The widely popular captive species in recent years are the colorful tropical forest dwellers. The desert cockroach family is the fourth largest family with fewer than 300 known species. These could also be called the hairy cockroaches since the pronotum always has a fringe of small hairs along the front edge. *Arenivaga, Eucorydia, Ergaula, Polyphaga*, and *Therea* are commonly kept genera. Females have a hump at the rear of the last visible sternite but they do not have notable valves for forming oothecae. At rest, the wings almost always fold

Ectobiidae: *Anallacta methanoides* nymph and adult

Blaberidae: *Macropanesthia rhinoceros*

without any fan-like pleating. In desert species the males are long winged and frail, while females are round, heavy, and wingless. The gender-biased difference in shape carries over to many forest species. The circular shape of the female may have led to many cases of beetle mimicry.

Polyphagidae: Adult female "desert" roaches (one female holding an ootheca)

Cryptocercus punctulatus: This is one of the young from the 2004 photo. They took five years to mature (Sept. 2009).

Cryptocercidae—There are a dozen or so large (around 30 mm) species that look nearly identical (even internally for most species). Males and females are difficult to tell apart. This is the only small family with any captive history, but the long life cycle and relative difficulty make captive stocks rare.

Nocticolidae—These are tiny, frail species from tropical Africa and Asia which inhabit caves, live under stones, or are found in termite nests. They lack true compound eyes and have clusters of ocelli or no eyes. They possess large cerci with relatively long setae.

Lamproblattidae—Ten South American species, including one from the genus *Eurycanthablatta*, comprise this family of rarely seen and ordinary looking, small species.

Tryonicidae—Less than a few dozen species from Australia, New Caledonia, and New Zealand make up this family, which had been a subfamily of the Blattidae.

Termitidae—There are traditionally twelve families of termites that can be placed in the epifamily Termitoidae.

MORPHOLOGY

The body plan of the Blattodea is so rudimentary that it is used in many books as the diagram of the basic insect body. Robot designers study cockroach morphology in hopes of constructing simple machines to follow the time-tested design. Roach mouthparts and overall design are pretty basic. Despite what seems to be the unchanging look of the cockroach body, there are a number of spectacular species with strange horn structures, wing variations, or irregular body shapes. There is some variability in the cockroach body, most notably in the

Tibial spikes on *G. oblongonota*
© Henry Kohler

Glass climbing tarsal pads on
G. oblongonota © Henry Kohler

volvating species, but variety is also seen in the legs, antennae, eyes, pronotum, and wings of many groups.

Legs are adapted to many purposes. Burrowing forelegs resembling those of mole crickets are found on species across unrelated genera including *Geoscaphaeus*, *Macropanesthia*, and *Polyphaga*. Forelegs adapted for digging or running may also be used to hold or carry food. Cockroach legs tend to be covered in greater numbers of spines than legs of other orthopteroids with the exception of large katydids. Thick and spine-covered legs are common for large and burrowing species. The middle and hind legs are nearly always long

M. rhinoceros: Forelegs are often
spiny and adapted for burrowing.

Disruptive coloration is seen on antennae of some species, as with this adult *Gyna capucina.*

Face of a cockroach

and used for running, while the front legs are shorter and can be used to carry away food (or dig). Species with shorter legs tend to move much more slowly. Any cockroach keeper knows most species are capable of jumping, but at least one has back legs resembling a grasshopper. Enlarged rear legs are found on the unusual *Saltoblattella montistabularis,* a tiny species from South Africa (Bohn et al. 2010). Just because other cockroaches do not have enlarged rear legs, should not be misconstrued to imagine they cannot jump skillfully. Another rear-leg modification is seen in the small *Eurycanthablatta pugionata* Fritzsche and Zompro, 2008, from Brazil. Males have a spine on each rear femur used in sparring over females like certain phasmids (*Eurycantha* and *Megaphasma*). Many types of roaches have their ears on the legs like orthopterans.

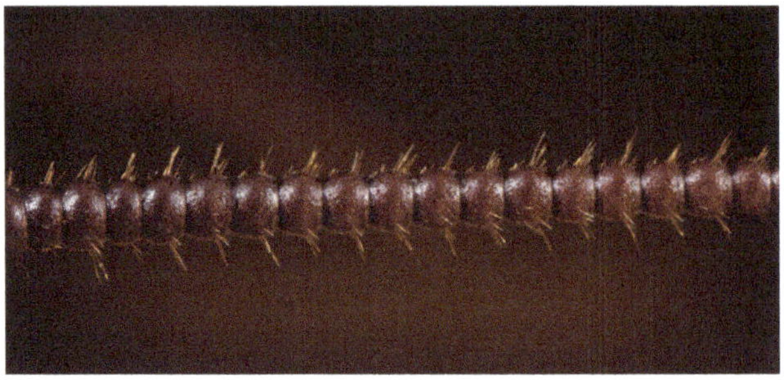

Turkistan cockroach antenna close up © Henry Kohler

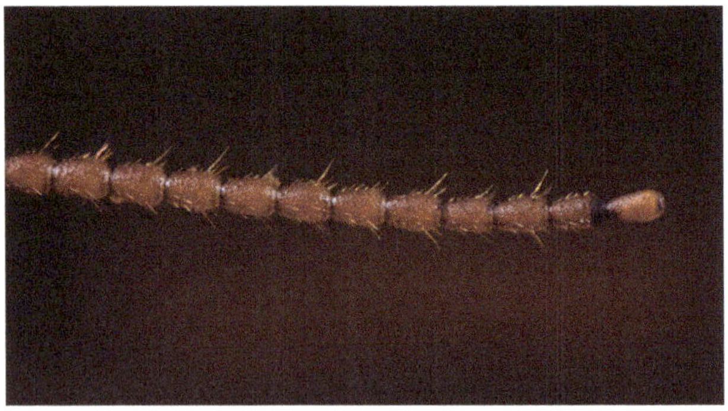

Turkistan cockroach antenna close up © Henry Kohler

Antennae of male *Gromphadorhina* are usually very fuzzy.

Cockroach antennae are as notable as the legs and are often longer than the cockroach itself. The antennae are flagellate rather than prehensile, simply meaning they move by muscles near the base and lack muscular joints further out. The antennae of young nymphs are shorter and comprised of fewer segments than adults. Though most are a solid color, there are a number of species with banded antenna (usually with just one or two pink or white bands). Antennae are groomed and cleaned with the mouthparts on a daily basis. The necessity to keep antennae clean for long-term use may be the reason most orthopteroid antennae never come in the intricate, feathery forms seen among the short-lived beetles, flies, moths. In some genera, including *Therea* and *Gromphadorhina,* the male's antennae are much longer and thicker than the female's. The sensory setae can be very large and give the male's antennae a fuzzy or furry appearance.

Many cockroaches have a pair of large compound eyes that wrap around each side of the head like the eyes of a dragonfly or mantis, but there are exceptions. Most of the burrowing

Annalacta face

polyphagid females have eyes that look like narrow stripes painted along each side of the head. Males of this same family can have eyes so large that they nearly touch on top of the head. The burrowing *Macropanesthia* species have small triangular eyes that make their faces look more like those of crickets than cockroaches. A few species, notably the Australian, cave-dwelling *Nocticola flabella* Roth, 1991, have no eyes at all. The compound eye is most often comprised of thousands of tiny facets. The individual ommatidia are too small to count readily under 10x magnification, even on the large eyes of *Blaberus giganteus*.

Polyphagid eye

In addition to the main pair of compound eyes, the lion's share of species have small

Blaberus giganteus ocelli

Turkestan cockroach: Even under magnification the facets of the cockroach eye are difficult to count. © Henry Kohler

Pronotum horn structure seen on an adult male *Macropanesthia rhinoceros* (adult female center, five year old female nymph right)

Turkistan cockroach ocelli close up © Henry Kohler

ocelli located between the antennae. One uncommon feature of cockroach ocelli is that there are only two of them. Most insects, including the very closely related mantids and phasmids, as well as bees and dragonflies, have three ocelli. The other major insect group with only two ocelli is the heteropteran section of the true bugs. A second unusual aspect of cockroach ocelli is that they often lack a raised, hemispherical lens and look like light-colored spots, rectangular lenses, or shallow indentations.

The cockroach pronotum is never much longer than it is wide, which is the principal visible difference from the praying mantis body shape. The pronotum can be variable in other ways. It can be hairy, spiny, bumpy, uniquely colored, or outfitted with strange, colorful organs. This body segment can be adapted into a shovel shape, flattened, folded backwards, or outfitted with horns. Horns are found on a variety of species and genera. In the Blaberidae there are various species across dozens of genera in which males have horns on the pronotum (not on the head). Horned species are usually blaberids, including the tropical New World *Hormetica* and *Lucihormetica* and Old World *Gromphadorhina*, *Macropanesthia*, and *Panesthia*. Except for *Panesthia*, these horns are greatly reduced or entirely lacking on females. The distantly related *Cryptocercus* have somewhat similar looking bumps on the pronotum and the size and presence is also not related to gender. Most species that live in loose dirt and leaf litter have a large, flattened pronotum used to plow through the substrate and protect the head.

Wings are highly variable across species and genera, and can even vary between siblings. The outer, first pair of toughened wings are called tegmina. Long tegmina usually protect membranous hind wings that are used for flying. Hind wings are lacking or severely reduced in many species, most often those with reduced tegmina. A large number of species lack wings altogether at maturity. In other species only the females are wingless. Wings tend to be similar in shape and size for members of the same species and gender, however there are some species, notably members of the genus *Byrsotria*, whose males can have longer or shorter tegmina between populations or, less often, in the same stock. Wing variation in cockroaches is very similar to variations seen in other orthopteroids.

At the end of the abdomen there is a pair of appendages known as cerci. These are sensory in function and are usually long and segmented. In many species they are useful in detecting air movement and therefore sound to

Large bug eyes

Horns on *E. javanica*

a limited degree. They act to protect the rear from attack and are longest in exposed species that lack the protection of substrate or tunnels. The cerci tend to be shortest on burrowing species, and some even retract into small grooves. The cerci of cockroaches are never incredibly long like mayfly (Ephemeroptera) cerci, nor adapted into pincers like earwigs (Dermaptera). A second pair of smaller appendages, called styles, is located near the cerci. Styles are found on most male cockroaches (and grasshoppers).

Common cockroaches are rather plain-looking creatures which come in varying shades of black and brown and barely demand notice in their natural habitat. There are some day-active species decorated with wild patterns, metallic tones, or contrasting red, green, blue, yellow, and orange pigments. Less often, species that are nocturnal or active only around dusk and dawn are also brightly marked. While the most colorful species are usually considered the most beautiful, the typical nocturnal brown roach can appear quite attractive with varying brown and tan markings.

In addition to standard coloration, some cockroaches fluoresce under ultraviolet light. Light-colored areas of the pronotum and tegmina of most species in the genera *Blaberus, Eublaberus, Hormetica*, and *Lucihormetica* (especially *B. giganteus*) glow a ghostly green. The compound eyes of large blaberids, specifically *Gromphadorhina, Hormetica, Lucihormetica*, and *Macropanesthia*, glow a strong green. Members of other families and all the small species of blaberids from genera including *Diploptera, Gyna, Nauphoeta, Oxyhaloa, Panchlora*, etc. do not possess glowing eyes (nor do *Blaberus*). As with many of the regular pigments, the exact purpose of the fluorescence is uncertain and it may have no utility. Scorpions and certain millipedes (famously the polydesmids) glow brightly under ultraviolet and there have been many suggestions as to why this may occur. The pair of spots on *Lucihormetica* have been said to protect these species through mimicry of bioluminescent click beetles. In captivity the spots do not glow except under UV, and the fluorescence of the pair of central spots is in the orange spectrum, whereas *Pyrophorus* click beetles glow green under their own power.

LONGEVITY

Cockroaches have a much wider range of longevity than their mantis relatives. (Praying mantids commonly live six to nine months from hatching to death, very rarely twice that.) A number of species do live just one season as adults, but they have a longer nymphal period because temperate species in the genera *Ectobius, Ischnoptera, Parcoblatta, Pseudomops*, etc. overwinter as nymphs (while the praying mantids only overwinter as oothecae in colder areas). Wood-feeding *Cryptocercus* also overwinter as nymphs in cold areas, but through multiple years. They live an average of seven years in total due mostly to slow development. The adults usually do not live more than two years. Tropical *Panesthia* can reverse these numbers with two years taken to reach maturity and adulthood lasting up to five (McMonigle and Willis 2000). Small tropical species like *Panchlora* and *Schultesia* can reach adulthood in less than six months, while even the longer-lived females rarely continue on more than another six. Males of other cockroaches live as little as thirty days (such as *Arenivaga* and *Corydidarum*), while females of the same species can live for seven hundred and thirty days. The record for documented longevity comes from an adult female *Macropanesthia* that lived seven years as an adult (Rugg and Rose 1991), and likely took four to five years to mature (at the very least three). Termite reproductives tend to be similarly long-lived, though some grow older. *Mastotermes darwiniensis* have survived in the laboratory for 21 years (Thorne et al. 2002).

LIFE CYCLE

Roaches are an ancient order of insects that arose before metamorphosis appeared and thus retain one of the least complex life cycles (incomplete). There is no pupa stage. The young develop inside eggs and hatch into fully mobile and active immatures. Nymphs appear similar to their parents but never have wings and may be colored or behave differently. Unlike groups with more complex life cycles such as flies, butterflies, and beetles, adult and nymph blattodean usually eat the same food and live in the same habitats. Nymphs and adults may even behave similarly, though nymphs of many species enjoy burrowing more than the adults do and (of course) never fly.

Cockroaches in the various families of the order Blattodea begin life in one of two basic ways. Either they hatch from thick-walled external oothecae or are born live. For live birth, bare eggs or membrane-walled oothecae are internally incubated and provided moisture or nourishment from the mother. Gestation consideration for species that incubate eggs internally is similar since the health of the mother is required for survival of the eggs. The husbandry requirements for egg-laying species are slightly different because the oothecae have separate requirements and should not be allowed to dry out or be kept too moist. If properly maintained, nymphs will emerge after a few weeks to nine months, mainly dependent on species.

There are usually seven to nine nymphal stages (instars) leading up to the adult reproductive stage. Males almost always average a smaller size, but require the same number of molts as females. This is not true for all species. *Diploptera punctata* males only molt three to four times, with three being most common in communal caging (Holbrook and Schal 1998). Oddly, *D. punctata* males display little variation in size, considering the total number of molts. There have been rumors of supernumerary molts for large *Gromphadorhina* individuals. However, small and large males that I have tracked through consecutive molts all had the same number of molts. Estimates based on measurements of wild *Macropanesthia rhinoceros* nymphs assumed ten to twelve instars (Rugg and Rose 1991), but an estimate is only a guess. It is a long-lived species with a number of molts—the only molt sequence documentation from birth to maturity showed five male and five female specimens all went through exactly nine molts, even though the time to maturity varied by more than a year and a half (McMonigle 2011b).

Molting takes place at the end of each instar. As nymphs grow, the outer shell does not grow, has limited ability to stretch, and becomes too small. A new exoskeleton begins to form beneath the old shell. Once the new exoskeleton is formed, chemicals are released that dissolve the lower layers of the old one. The animal may look pale or gray as this occurs. A few days later a pre-existing seam across the back splits open to reveal a larger animal underneath. The exoskeleton is puffed up with air and then deflates about twenty minutes later as it hardens. At first the emerging roach is bright white, except for two deep black eyes. The new exoskeleton transforms over the next few hours to normal coloration and rigidity. Before it hardens and darkens, the exoskeleton is termed *teneral*. A hormone produced by living animals called bursicon aids in the tanning (sclerotization) process. If the animal dies, it

The elusive white cockroach is always recently molted (*Gromphadorhina oblongonota*).

Developing color after molt *(Aeluropoda insignis)*

Growth in *Macropanesthia rhinoceros*

Growth stages, newborn to adult, in *Gromphadorhina oblongonota*

will never fully harden (sclerotize). This chemical hardening process takes place under water for many arthropods and the exoskeleton hardening has nothing to do with air or "drying out." Water cannot soften an exoskeleton, but it is an important lubricant needed for shedding. Air and fluid are required to properly inflate the new shell. After the molt is completed the abdomen becomes broad and shortened, but the roach eats over the next few days and weeks to grow plump and elongate again. Once they reach adulthood (often indicated by short or long wings) they never molt again. There is never a post-ultimate molt (unlike females of the widespread pet tarantulas).

The newly molted cockroach gives rise to the myth of the elusive 'albino' cockroach. So far a true albino cockroach has never been documented. (An albino is an aberrant form lacking normal pigment.) Early instar exoskeletons can be naturally transparent. There is one completely white and eyeless *Nocticola* sp. from Australian caves, but it is not an albino because it is naturally white. There are quite a few pale colored cave species, but a complete lack of pigment is rare. There are also some very pale forest-dwelling *Panchlora* adults and sand-dwelling *Arenivaga* males, but these too are naturally occurring, and have small areas of dark (melanin) pigment. I have seen a single pied (incomplete albino) individual of *Blaptica dubia* but it was deformed. It is possible the deformity prevented circulation of the bursicon hormone rather than being a case of true partial albinism.

GENDER DIFFERENCES

Sexual dimorphism and sexual dichromatism can be very pronounced in adult cockroaches. Skinny, long-winged males often mirror rotund, short-winged (or wingless) females. Nymphs of both genders start out nearly identical but can usually be separated as they develop, long before reaching maturity. Gender differences in longevity and morphological differences in standard body parts were detailed earlier. Some dimorphisms are more certain but less obvious.

Size may be the most familiar dimorphism. From above, a noticeably larger size identifies the females of *Blaberus colosseus, Opisthoplatia orientalis*, and *Panchlora* species, even though genders have essentially the same wing structure and coloration. *Blattella, Gyna,* and *Schultesia* can almost always be picked out by size, but not with certainty. In the vast majority of other genera lacking extreme dimorphisms the females tend to be larger, but it is not rare for them to be smaller than males.

Various egg-laying genera including *Blatta, Deropeltis, Parcoblatta*, and *Polyphaga* are sexually dimorphic as adults and are easily sexed by body shape and wings. Determining the gender of other egg layers like *Blattella, Cryptocercus, Drymaplaneta*, *Eurycotis, Neostylopyga, Periplaneta, Pseudomops,* and *Therea* can be very tricky at a glance. Upon closer examination, sexing these egg layers is made easy by comparing the last visible lower abdominal segment, known as the terminal or anal sternite. The last visible sternite of the male is the narrowest and smallest segment. Females possess a very large terminal sternite which is really the second to last sternite covering the last. In female Blattidae (*Blatta, Deropeltis, Periplaneta, Neostylopyga*, etc.) this segment is followed by very noticeable smaller, paired appendages used to form the oothecae. These segments or 'valves' are the upper and lower terminal segments divided in half. Many polyphagid females have a depression or bubble in the terminal segment that might be mistaken for a small segment, but there is no suture.

Male abdominal segments *(Macropanesthia rhinoceros)*

Female abdominal segments *(Macropanesthia rhinoceros)*

Female *Macropanesthia* scoop and pronotum shape

Macropanesthia male scoop and pronotum shape. This minor adult male does not have the points (horns) of a major male but it still has a very different shape than the female.

Female *Blaberus giganteus* abdominal segments

Male *Blaberus giganteus* abdominal segments

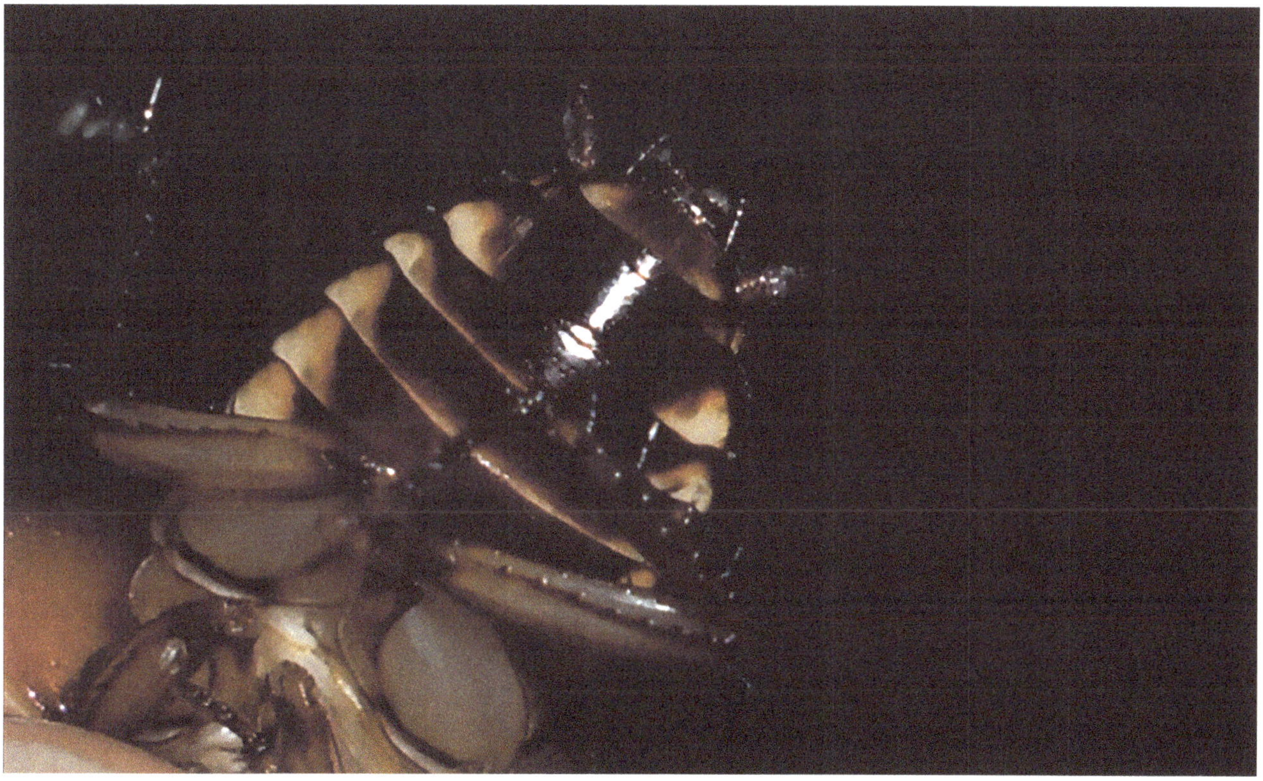

Male *Eurycotis decipiens*

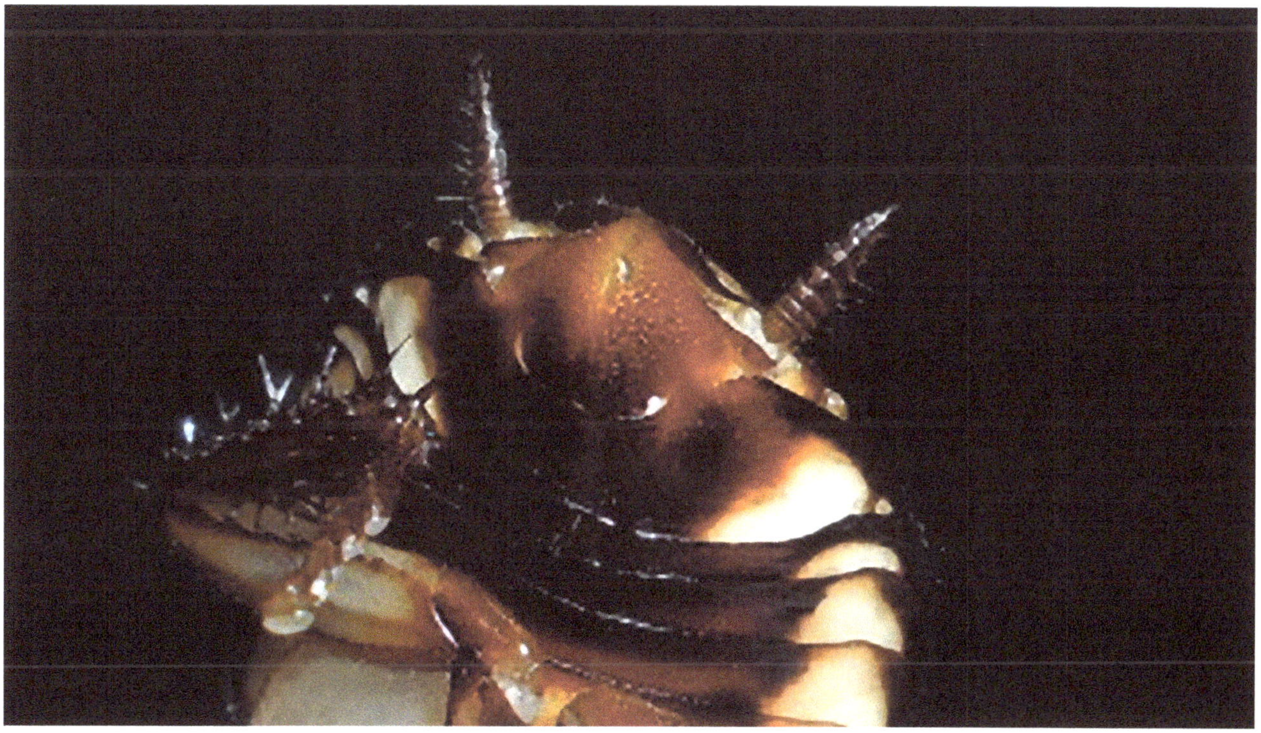

Female *Eurycotis decipiens*

Even short-winged species can be dimorphic (male at right).

Dimorphism in *Therea*: female (left) has a large notch in pronotum and thinner, shorter antennae.

Some live-bearing species can be sexed by the shape of the pronotum. Adult males of various genera (for example *Aeluropoda, Elliptorhina, Geoscapheus, Gromphadorhina, Hormetica, Leozehntnera, Lucihormetica, Macropanesthia,* and *Panesthia*) can be sexed by the presence of horns, notches, or scoops at the front of the adult male's pronotum. These structures are usually used for fighting over mates or territory. In the various glowspot cockroaches (*Lucihormetica* species), the males also possess a pair of bulging, relatively soft organs on the pronotum among the horn structure. The gender of other live-bearing genera (*Corydidarum, Byrsotria, Pycnoscelus femapterus*, etc.) is easy to determine at a glance because the females are more rounded and have short wings or none at all.

The most accurate method for determining blaberid gender is similar to sexing of egg-laying roaches. The terminal lower abdominal segment or sternite visible on the female is the largest segment. (Live-bearing females, however, never possess conspicuous appendages for egg case formation.) The last visible segment on the male's abdomen is the narrowest and smallest. Members of the subfamily Panesthiinae and Geoscapheinae are also sexable in this manner, but the last visible segment on the male is extremely tiny. Some genera can be sexed by looking at the terminal sternite early on, although young nymphs of *Blaberus* and many other genera appear to be all males. Many species appear to be 100% males for the first few instars because the last segment is covered over by the second to last segment only as the female develops. In this case, nymphs can be sexed after the initial instars have been passed (usually on or before the 4th).

Mature males of most cockroach species have tiny appendages near the cerci known as styles. The presence of styles on a sexually mature specimen is usually a clear indicator of gender. Nevertheless, these appendages are tiny even on large species and can be difficult to see or hidden (rarely nonexistent).

CRYPSIS, MIMICRY, AND OTHER DEFENSES

Many members of the Blattodea would not be identifiable as a cockroach by a person lacking basic familiarity with entomology. Some species would require even an experienced entomologist to double-check their identification. While many roaches can be difficult to see among forest litter, others imitate overactive wasps or strikingly colored beetles. Mimicry in the Blattodea ranges from minimally impressive to incredibly accurate and helps to protect these animals from being eaten in the wild.

Numerous cockroaches live among dead leaves and rotting wood on the forest floor. A great number of species have texture and coloration adaptations that help them to blend into their surroundings. One commonly kept species, the Madeira cockroach (*Rhyparobia maderae*) has been referred to as the 'dead leaf roach' since its mat brown outer wings with dark veins make it difficult to discern among leaf litter. Some of the *Rhabdoblatta* have been referred to as 'leaf' roaches, but they do not have notable vein markings seen on other orthopteroid leaf mimics. The supposed leaf mimics pale in comparison to the crypsis of nymphs of *Opisthoplatia* and *Rhabdoblatta,* which blend perfectly into the textured surface of wood. The exoskeletal structure can allow animals to blend into substrates in less obvious ways. Many nymphs have short setae or roughened textures on the surface of the thoracic and abdominal segments. The structures gather particles of dirt as the animals move through their habitat to provide camouflage (though none are known to deliberately collect

The surface of these *Blaberus giganteus* nymphs seems to be built to hold materials which can aid in camouflage.

items and glue them to the exoskeleton like decorator crabs or certain assassin bugs). Lastly, some nymphs change from light brown to dark brown (like most tree bark and decayed leaves) in the presence of excessive moisture. It does not help to match substrate that darkens following a rain if your disguise cannot follow along. There is a double layer of cuticle separated by air. The color change occurs when the air just beneath the outer permeable layer of the cuticle is replaced with water.

A dirt-covered exoskeletal surface can serve to camouflage nymphs. Specimen on right was washed off (*Byrsotria cabreri*).

Panchlora nivea is conspicuous in a cage but cryptic among green leaves.

There are species that live among the leaves of live plants and possess cryptic green coloration. Dozens of *Panchlora* species found throughout the New World tropics come in various shades of green. The most widespread, the green banana cockroach *Panchlora nivea,* is so named because it has been found as a stow-a-way among similarly colored, unripe bananas. Due to their bright green coloration and great agility, catching these cryptic little critters among live plants is practically impossible. However, *Panchlora* nymphs escape predators in a different way. They are dark brown and live buried in the ground. Other genera, such as the egg-laying *Mediastinia* from Australia, can be green as nymphs and adults. They live arboreal lives among ginger and palm leaves (Rentz 2014).

A number of small roaches are found on flowers but, unlike some Mantodea, none of them look like flowers. Species found during the day on flowers include warningly colored *Pseudomops septentrionalis* in North America (McMonigle 2007c), brightly colored *Ellipsidion* species in Australia (Rentz 2014), and beautiful metallic-blue and orange *Eurcorydia* like the Taiwanese flower cockroach (Zha 2015). Species usually found about flowers tend to be warningly colored or mimic wasps, though the tiny *Ectobius* spp. can be found on flower stalks and are rather drab. There is a very small, pale species known as the orchid cockroach, *Shelfordina orchidae*, but it does not look anything like a flower. It feeds on the flower petals and root tips and can be a pest of orchids in greenhouses.

A few small species imitate insects with painful stings or bites. Most invertebrate predators discover early on the importance of avoiding the painful sting of wasps. Some of the *Pseudomops* and *Pseudophyllodromia* are day-active like wasps and sit in full view on leaves and flowers. Not only do they mimic size, color,

Opisthoplatia orientalis nymphs can be very difficult to see on wood or bark.

The adult color of *Opisthoplatia orientalis* is very bright and certainly not cryptic.

and shape, but also copy the wasp's speed and jerky, "nervous" body movements. They also can have antennae banded in black and yellow like wasps. Species in the genus *Paranauphoeta* are assassin bug (Hemiptera: Reduviidae) mimics. They have the bright spots and appearance of the true bugs, famous for their painful bites. Only the adults seem to mimic assassin bugs, since the cockroach nymphs have an entirely different color scheme. A few species mimic ants but probably not stinging ants. Bell et al. (2007) say the Indonesian nocticolid, *Cardacopsis shelfordi*, mimics an ant. The hatchlings (only) of some commonly reared species like *Eurycotis decipiens* and *Hemithrysocera* spp. also seem to resemble ants.

A number of brightly colored species have their own warning colors and likely do not mimic anything. *Deropeltis paulinoi* are protected by a quinone secretion and are black and orange in color. *Eurycotis decipiens* advertises its powerful defensive odor with black and white stripes. In Costa Rica *Euphyllodromia* sp., tiny day-active cockroaches with bright white- and black-striped bodies, can be seen, though adults (possibly wasp mimics) are not nearly so garish. Sometimes only the adults are colorful while the nymphs are cryptic and brown, as is the case for *Opisthoplatia orientalis* and *Pseudomops septentrionalis*. They do not seem to have chemical protection and it would be difficult to imagine they model a specific creature, but the white, black, and red are standard warning colors. The nymphs of *Henshoutedenia* and *Simandoa* have contrasting banding while adults have gray wings and look very different.

The most accurate and impressive mimics are certainly roaches that mimic beetles. Although beetle larvae look incredibly different from nymphal roaches, the adult body plan is not very different. Unlike most cockroaches, even nonpoisonous beetles have a hard shell and limited food value. The Pacific beetle mimic roach *Diploptera punctata* looks just like a small brown beetle and even folds its wings so that it appears to have a straight line down the middle (like the line between the elytra of a beetle). Its strong defensive secretion smells similar to the quinone defense of darkling beetles (Tenebrionidae). Adult females of *Ergaula* species also seem to mimic brown beetles, though they lack the powerful quinone defense. *Paratropes lycoides* and other *Paratropes* species look identical to Lycid beetles, which are foul tasting and sport bright warning coloration. Birds remember the color and foul taste, which deters them from eating either beetles or cockroaches. *Therea petiveriana* has a body shape nearly indistinguishable from an adult tortoise beetle (family Chrysomelidae) or carrion beetle (family Silphidae), though the warning coloration of black with white spots is only familiar on Indian scarabs (family Scarabaeidae) and tiger beetles (family Carabidae) with quite different body shapes. They do not have a strong chemical defense, but they do have eversible organs, which broadcast a smell of rotting flesh when the animal is distressed. The firefly cockroach *Schultesia lampyridiformis* is named for its resemblance to adult fireflies (Lampyridae). Few animals enjoy the foul taste of fireflies, but *S. lampyridiformis* are commonly used as feeder insects. Tiny *Prosoplecta* species from tropical Asia are fantastic and accurate mimics of ladybugs.

Although roaches resemble many things, mimicry is not the reason for every resemblance. Pillbug cockroach females (*Corydidarum pygmaea*) are the same shape and size as pillbugs, terrestrial isopods that roll into balls when threatened. It is likely the volvating

These pale brown Ectobiidae are resting in curled leaves in a Philippine rainforest, Panay. © Christian Schwarz

Firefly mimics, *S. lampyridiformis*

form required for this physical defense is the only reason for the visual likeness. Some cockroaches are referred to as limpet, horseshoe crab, and trilobite cockroaches, but these unrelated gastropods and arthropods are marine and could never afford a cockroach defense by association. Roaches that loosely resemble limpets simply have the same physical defense (holding fast to a flat surface and offering no edge to bite onto). The resemblance to trilobites is more of a loose coincidence than a case for convergent evolution. Trilobites were mostly round, flattened, and multi-segmented arthropods, but they were no closer to a cockroach than a shrimp or scorpion. In Australia all the species in the genus *Laxta* are known as trilobite roaches (Rentz 2014) (*Laxta* look a lot like South American *Blaberus* relatives). In the United States various cultured *Byrsotria, Hemiblabera*, and nymphs of *Blaberus* have been tagged as trilobite, woodlouse, or horseshoe crab roaches over the years, with none holding the common name for long. I have not seen a roach that really looks much like a real horseshoe crab, but the name is especially fun since the roaches are named after something, named after something else (this marine arthropod is related to arachnids and is not a true crab).

Cockroaches that look like other animals are a favorite among enthusiasts. The mimics are among the most beautiful of all the roaches and a few are arguably among the most beautiful of all the insects. Although fireflies, wasps, chrysomelid and lycid beetles are generally very difficult and time-consuming to rear, the roaches that look like them, fortunately, can be as easy to maintain in culture as any cockroach.

CHEMICAL, PHYSICAL AND AUDITORY DEFENSES

Many blattodeans have more than one defense against being eaten. Even the cryptics and mimics mentioned above have a variety of chemical or physical tricks to back up their visual defense when it fails. Visual defenses usually are designed to work on vertebrate predators while there are a host of invertebrate predators likely to pose greater risk to roaches.

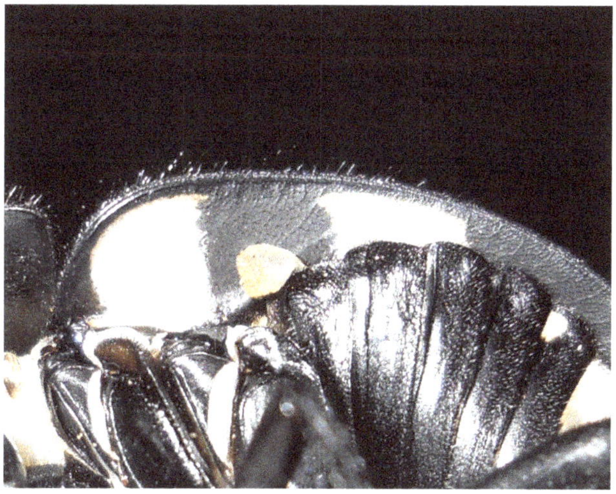

Eversible odor gland of *Therea petiveriana* smells weakly of rotting flesh.

Odor organs are common and activated when the animal is greatly distressed. The variety of smells can be a little different for each species, but is usually similar within genera. *Gyna lurida* smell like snake urine when disturbed, while the odor from *Ergaula pilosa* is more like rotting flesh. A number of other polyphagids, including the beautiful *Therea*, have eversible defensive organs which can expel a pungent rotting flesh smell. *Blaberus* and *Eublaberus* produce a smell similar to mammal urine when distressed. *Eurycotis* species produce very strong defensive odors, but the result is a pungent chemical smell that does not resemble animal stench. Odors can be used to communicate with other roaches in addition to or instead of defense. The same secretions can be of enough volume to produce visible droplets, blurring the lines between odiferous and liquid defenses.

Liquid defensive secretions range from smelly and foul to seemingly innocuous. The

Eversible organs contain odiferous chemicals and are turned inside out when the animal is distressed (*Ergaula pilosa* male).

Many species spit in self defense (*Lucihormetica verrucosa*).

liquid defense of *Deropeltis* and *Diploptera* comes out as burning droplets containing quinones (an effective defense also used by many millipedes and darkling beetles). Even the strongest chemical defenses tend to be harmless under normal circumstances to the animal they protect, though the fluids of *Pelmatosilpha larifuga* have been reported to kill the cockroach that produces them (Mann 1995). *Rhyparobia* and other Nauphoetini use their back legs to flick smelly liquid excreted from their anus. A number of nymphs and small species, including African bullet roaches, exude clear, odorless, sticky liquid that covers the rear abdominal segment and can protect them from ants. This fluid is present under normal circumstances and is not created as a result of a disturbance. Under distress many large-bodied genera, including *Gromphadorhina*, *Blaberus*, and *Lucihormetica*, will spit large droplets of clear, odorless fluid. Others like *Therea* spit orange or brown smelly fluids from the mouth like grasshoppers.

The most familiar physical defense of the cockroach is its ability to run (or fly) with lightning speed to the closest hiding spot. People often talk about how pest roaches run for cover with unreal speed when the lights are flicked on. Most species, even many brightly colored or cryptic ones, hide during the day and come out only when it starts to get dark. The flattened shape of many species aids in escape and concealment.

Eurycotis lixa: In this species the chemical defense covering the rear segments is later spread over the whole body.

Eurycotis lixa: A protective coating can backfire since dirt sticks to it, but it is not a mechanism for camouflage.

Spiny legs offer some physical protection and are most common in large species. It is very difficult to hold a *Blaberus* or *Gromphadorhina* that does not want to be held. *Blaberus* may also try to bite, but the bite is relatively powerless. (Only *Macropanesthia* might offer a good nip, but not in defense, rather when fingers smell of dog food.) Spines are often less numerous and shorter in small species, since they cannot provide much power to them. Small species tend to employ mats of hair-like setae as a physical barrier rather than an active defense. However, small species can employ defensive spines. Adult *Panchlora nivea* have tiny spines that break off at the end and can cause surprisingly uncomfortable, if short-lived, pain to the human hand.

This *Elliptorhina chopardi* is spitting . . .

. . . and defecating in self defense.

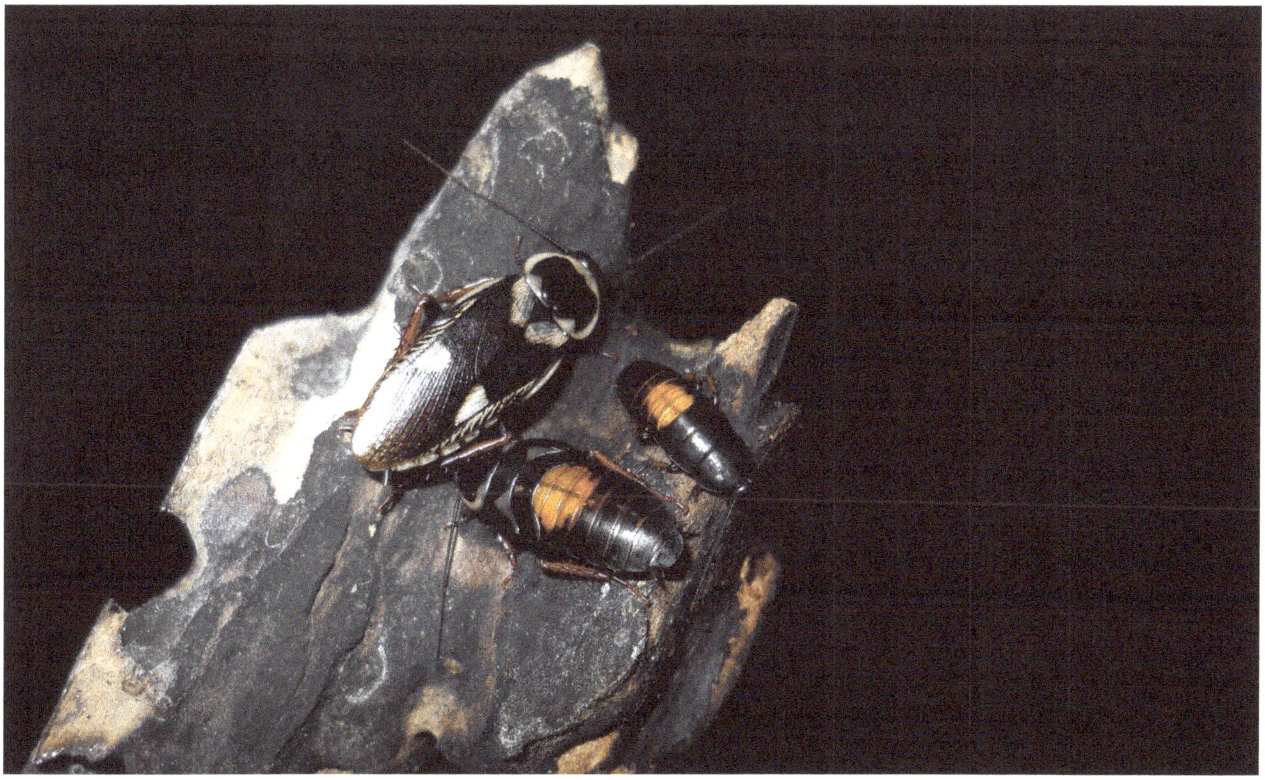

Warning coloration in *Paranauphoeta*

A very thick, hard exoskeleton is the ultimate defense.

Warning colors can be somewhat subtle (*Oxyhaloa duesta*).

Warning coloration in *Pseudomops*

In some species like the extremely tiny *Euthlastoblatta gemma,* bright colors may be useful for disruptive patterns, as a form of crypsis. This one was found in Opelika, AL, under the bark of a standing tree. © Alan Jeon

Autotomy (shedding a body part to escape, like a lizard dropping its tail) is common, though some of the discarded body parts never regrow. Many of the egg layers like *Periplaneta* shed legs with reckless abandon when handled. Legs can grow back if there are enough molts to go, but substantial regeneration usually takes at least three molts. *Eurycotis decipiens* is infamous for shedding its wingpads during handling. Of course these will not grow back since adults do not molt.

Sound production as a defensive mechanism is well known in cockroaches. The most famous example comes from the popular hissing cockroach, *Gromphadorhina portentosa*. These animals, especially adult males, rapidly expel air through the spiracles resulting in a hissing sound. Various related species produce sounds, including all known species of *Gromphadorhina*, *Elliptorhina*, and *Aeluropoda*. Distantly related species such as *Macropanesthia rhinoceros* are capable of making a similar, if less impressive noise. *Blaberus* also make raspy hisses by rapidly contracting and expanding the abdomen. A number of species only make noises at maturity since the tegmina (outer wings) are used to generate sound. *Nauphoeta, Rhyparobia, Simandoa,* and *Henshoutedenia* adults make noises by rubbing the pronotum against the tegmina. Possibly the most uncommon method of sound production is found in species of the large South American *Megaloblatta*. The abdominal segments of the nymphs have sound-producing structures that make a shrill constant hiss or rattlesnake noise depending on how the abdomen is rapidly shaken from side to side (Schal et al. 1982).

Autotomy of the legs is very common for blattids and ectobiids. The normal point of breakage is the base of the femur.

Autotomy of the tegmina is an unusual defense (*Eurycotis decipiens*).

This black armored cockroach (*Catara rugosicollis*) from Singapore (adult female 20mm+) is capable of producing odiferous liquids that can be difficult to wash off skin. © Melvyn Yeo

Dirt-covered immature specimen in Trinidad © Martin Gamache

An assortment of Blattodea from the University of Texas Insect Collection for the “Insects Unlocked” project at the University of Texas at Austin (arrangement by Julia Suits, photographed by Alex Wild)

It is possible to tell the animal with the margined pronotum on the right is the male because his wings are below the females during mating. Ectobiidae in a Panay rainforest. © Christian Schwarz

Macropanesthia rhinoceros mating pair (male on left)

REPRODUCTION 繁殖

The way in which cockroaches perform courtship has been extensively covered in literature, if only for a few dozen species, primarily the pests (e. g. Roth and Willis 1954, Willis et al. 1958, Bell 1981). Females usually broadcast pheromones that attract mates from afar while other substances or courtship movements take over when they are in proximity. Antennae touching and movements are normal for almost

Eucorydia aenea dasytoides mating pair © Chen Zha

every species. Some females will begin to broadcast their receptiveness before the final molt. Frenzied males are often seen circling a female *Diploptera punctata* before she has fully removed herself from the old exoskeleton. *Macropanesthia* battle even more intensely than *Gromphadorhina* over mates, but are different because precoital courtship seems lacking and, more surprisingly, a male will remain with a female for up to a year before her final molt (McMonigle 2006a). In hissing roaches, the males battle over females with loud hisses but produce soft, short hisses during courtship (Barth 1968). In quite a few species the males have organs on the back, beneath the tegmina, which produce a substance dubbed 'seducin'. Differences in these tergal glands (located on the tergites) are used to identify some species. The male lifts his wings and the female feeds before she accepts his advances.

Mating pair of *Gyna lurida*, male at bottom

Blaberus giganteus mating (female at right recently molted and is still very pale)

Teneral female *Blaberus craniifer* (male on the right)

Diploptera punctata: an unconcerned male pins a teneral female in place during mating while cage mates consume her alive.

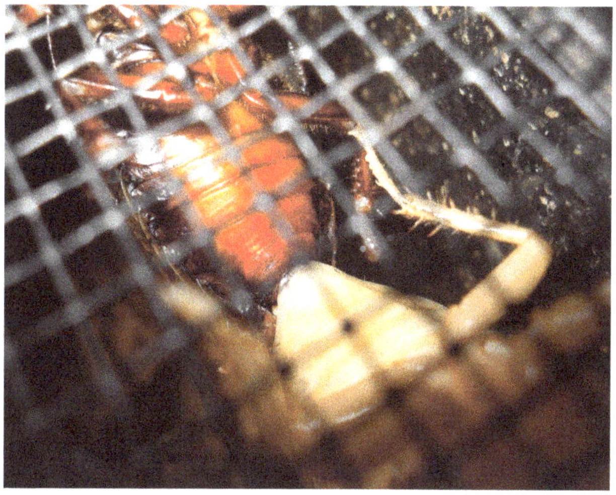

Teneral mating in *Eublaberus posticus:* the female`s wings were damaged as they hung down from the lid and the male kept her from righting herself.

Teneral mating in *Schultesia lampyridiformis*

Mating in cockroaches is not commonly observed (when looking in an enclosure it is far more likely not to see mating than to see it) compared to mating for the average beetle, cricket, or phasmid that just cannot get enough. This is because female cockroaches become unreceptive after a successful mating and are unlikely to mate more than a few times in their lives. The first male to succeed is the winner because his spermatophore prevents further matings. It may obstruct further entry to the spermatheca (the female's compartment that temporarily holds the spermatophore), but usually the female simply is unwilling to mate again. For a time scientists considered teneral mating uncommon and possibly advantageous to females although there was no evidence the male offered any chemical protection to the teneral female (Wyttenbach and Eisner 2001). However, teneral mating is a widespread behavior relating to mate selection because subsequent males are rejected (McMonigle 2012d). Teneral females are also at a disadvantage because the male pins them in place when they are the most vulnerable. In captive situations this often leads to damage or death as a result of cannibalism by other animals in the enclosure, sometimes even if tank mates are not at all hungry or thirsty. Similarly increased risk is likely encountered in nature from predators and colony mates, but reproduction is almost always more important than safety.

Mating dance charting may be academic, but there are a few ways the specifics of mating behavior can be telling for culture requirements. If a female cockroach performs the

This *Eurycotis opaca* forms an ootheca which she will deposit shortly after its casing hardens.

This *Simandoa conserfariam* forms an ootheca. When completed, it will be flipped sideways, drawn back into the abdomen, and safely protected from predators or cage mates until hatching.

Cockroach ootheca, Sarawak, Malaysia (CC-SA) Bernard Dupont

Hissing cockroach with offspring (CC-SA) Matt Reinbold

In some Ectobiidae species like this Malaysian "little Penquin" the ootheca never fully darkens and is carried by the female until just before hatching.

This (Blaberidae) *Schultesia lampyridiformis* female is forming the usual ootheca with eggs pointing up and down.

Once completed, the ootheca is rotated sideways (eggs pointing side to side) and drawn back into the abdomen.

characteristic mating call (arching of the abdomen as also seen in a female mantis) the male provided is almost certainly inadequate. This is uncommon, but not so rare if only one male is provided. If there are a limited number of females available they should be allowed to mature separately from the males to ensure their safety.

ON REPRODUCTION AND PHYLOGENY

LOUIS M. ROTH
Museum of Comparative Zoology
Harvard University, Cambridge, MA

"Ogni scarrafone è bello
a mamma sua."
(Every cockroach is beautiful
to its mother.)
—Neapolitan folk saying (Schweid 1999)

The fossil record shows that long before the dinosaurs, cockroaches were present during the Carboniferous Period (350 million to 400 million years ago). At that time they were so numerous in the coal seams, that the period was known as "The Age of Cockroaches." Since then thousands of animal species have become extinct, and it is obvious that cockroaches are survivors and can adapt to almost all ecological environments; they have been found on mountains (one species is even known from the Himalayas), in deserts, caves, animal burrows, and water. Fortunately only a handful of the 4,000 to 5,000 species have adapted to man-made structures and have become synanthropic, living as domiciliary pests with man.

Ninety percent of fossil cockroaches consist of whole or fragmentary tegmina, and only 5% are entire insects. Fossil cockroaches resemble present day species, but many from the Carboniferous and Permian deposits (270 to 225 million years ago) of Europe, Asia, and North America, as well as the Triassic (225 and 180 million years ago) and Jurassic deposits (135 million to 100 million years ago) of Russia had long external ovipositor valves, and these undoubtedly deposited single eggs ". . . in soil, organic debris, or soft plant tissue." (Vishniakova 1968, Vrsansky 1997). Over time the external ovipositor gradually decreased in length during the Jurassic Period and presumably evolved into the present day valves found in all extant species.

An important factor which enabled cockroaches to survive these millions of years was the females' ability to alter their oviposition behavior which protected her eggs from dangers inherent in the environment (i.e., desiccation and predators). A large number of cockroaches are oviparous or ovoviviparous, and only one viviparous species is known. Cockroaches were originally oviparous, and ovoviviparity evolved from oviparous ancestors. The steps by which this has occurred can be inferred if one keeps large cultures of several representative species so that their oviposition behavior can be observed. Differences in the oothecal membrane that surrounds the eggs can be determined.

Cockroaches have two ovaries consisting of various numbers of ovarioles which determine the number of eggs that will be formed in an ootheca. (The ootheca refers to both an outer membrane and the enclosed eggs.) Some species have as few as 4 or 5 eggs (*Alluaudellina cavernicola*), whereas others may have many more. Species of *Gyna* may have from 94 to 243 eggs in an egg case (Roth and Willis 1957b). After mating, the basal oocyte of each ovariole increases in yolk content, the mature eggs are expelled through a blob of soft, rubbery, colleterial gland secretion and, with the help of the ovipositor valves, are lined up vertically and

This female *Perisphaerus* in a tropical rainforest has young attached securely to her underside (Panay, Philippines). © Christian Schwarz

A male *Perisphaerus* sp., has a very different body (Panay, Philippines). © Christian Schwarz

When disturbed the *Perisphaerus* female rolls into a ball with her young safely protected. © Christian Schwarz

Another view of the *Perisphaerus* female with young in a Panay rainforest. © Christian Schwarz

alternately on one side and then the other, so that upon its completion there are two rows of eggs covered by a membrane. The membrane is formed from secretions produced by a pair of colleterial glands. All cockroaches that produce an ootheca extrude their eggs with the micropylar ends and oothecal keel directed upwards. The position in which the ootheca is carried at the time it is deposited (i.e., whether or not it is rotated 90°) is significant taxonomically and played an important role in the evolution of blaberid and blattellid ovoviviparity.

The following are the three primary kinds of cockroach reproduction:

1. Oviparity

The ootheca is dropped shortly after its formation; initially the eggs have enough yolk to complete development but needed water is obtained from the substrate; some species have a membrane which prevents water loss. The females may dig or chew a hole in the substrate, deposit the egg case in it, sometimes apply sticky saliva, and finally cover it with substrate particles.

In some oviparous species, the ootheca is carried in the original vertical position during gestation, or deposited from the vertical position (e.g., Blattidae and Ectobiidae: Pseudophyllodromiinae). In the Ectobiidae (Blattellinae), the ootheca is rotated 90° and then is usually deposited on a substrate, or carried in the horizontal position during gestation (e.g., species of *Blattella*). Those species that carry their oothecae externally during gestation have enough yolk to complete development but obtain needed water from the mother. The females of those species that deposit (and often hide) the ootheca just after it is formed have nothing more to do with it or the young that eventually hatch. When the eggs have completed development, the nymphs swallow air and expand; their increase in size and combined efforts separates the two halves of the oothecal keel which is a line of weakness, allowing the nymphs to escape the confines of the egg case. If all the eggs develop successfully, the hatch is usually ideal and all of the nymphs escape simultaneously. The species that deposit their oothecae shortly after they are formed, produce a greater number of egg cases, and more frequently than species that carry them internally or externally during gestation; the latter species, e.g., *Blattella* spp., behave like ovoviviparous species in that their ovaries do not develop during pregnancy.

A number of parasites attack cockroach eggs, usually inserting one or more eggs through the deposited ootheca, or while it is still attached to the ovipositing female. Evaniid wasp larvae have been found in the oothecae of two species of *Blattella* that were carried by the females.

2. Ovoviviparity

Except for the Geoscapheinae, all of the Blaberidae rotate their oothecae 90° and then retract the ootheca into the uterus where it remains until birth. As the eggs develop during embryogenesis they increase in size and by the time of birth they occupy practically the whole abdomen. Rotation of the ootheca reorients the ootheca so that the long axes of the eggs lay in the plane of the cockroach's width which makes it easier for the uterus to stretch once the egg case is retracted and the eggs increase in size.

Ovoviviparity also evolved in the Ectobiidae. In the Blattellinae there is an African genus, *Stayella*, whose ootheca is virtually identical with that made by *Blattella* spp., but it is rotated and retracted into the uterus where it remains until birth. In the Pseudophyllodromiinae, *Lophoblatta* spp. carry their oothecae in the vertical position during gestation;

their egg case is unusual in being wider than long and apparently this type of ootheca allowed the evolution of ovoviviparity in this subfamily (e.g., in *Sliferia* spp., *Pseudobalta* spp.).

In most ovoviviparous species, and the only viviparous species, the first instar nymphs are extruded and disperse at birth. But in some species the young remain under the body of the mother for a short time. However, in *Thorax porcellana* (Saussure) from India, the tegmina are strongly arched and dome-shaped, and after birth, the abdominal tergites become depressed forming a trough where 32-40 first instar nymphs, which have eyes, stay with the female for two instars; apparently, the nymphs puncture the intersegmental membranes of the mother and feed on her body fluids that ooze through the resulting pores. Similar behavior occurs in *Phlebonotus pallens* from Ceylon whose tegmina are arched, the wings greatly reduced, the upper surface of the abdomen is depressed and the sides are raised forming a chamber in which more than a dozen new-born young are carried about; whether or not the nymphs puncture the intersegmental membranes and feed on oozing fluids is unknown. In *Perisphaerus* spp. the first and second instars, which are blind and have very unusual elongate, non-chewing mouthparts, cling to the undersurface of the female and apparently obtain nourishment from her, possibly through a pair of openings between the mid and hind coxae; if attacked (e.g., by ants) the pill-bug-like female, rolls herself into a tight ball, with the young still clinging to her, and she and her young are safe from attack.

No hymenopterous parasites have been found in the eggs of Blaberidae. However, it is possible that they exist, and eventually will be discovered because evaniids have been reported in *Blattella* spp., showing that an egg case can be parasitized while it is being carried by the female. It is conceivable that during the period the ootheca is being formed by a blaberid, a wasp could deposit its egg(s) into the cockroach eggs, which would then be retracted into the uterus. Although it is conceivable that internal incubation of the eggs protects them from parasites, it should be pointed out that if a pregnant blaberid female is killed by a predator, her embryos suffer the same fate.

3. Viviparity

Only one cockroach, *Diploptera punctata* (Eschscholtz) is viviparous. The female produces a very small ootheca which is rotated and retracted into the uterus. Initially the eggs do not have sufficient yolk and water to complete development; the embryos obtain solid nourishment and water from the mother. There are only about a dozen eggs in an ootheca and the embryos compete for nourishment in the uterus; if the number of eggs are reduced experimentally, e.g., by removing one ovary prior to oviposition, the size of the smaller number of new-born nymphs are much larger than the greater number of control nymphs.

The evolution of ovoviviparity and viviparity in cockroaches necessitated a change from a hard, rigid, thick ootheca that was deposited shortly after its formation, to a soft, flexible, thin egg case which was retracted internally into a uterus. Such changes have taken place in the superfamily Blaberoidea (Ectobiidae, Blaberidae) but not the Blattoidea (Blattidae, Cryptocercidae, Termitoidae). In the Corydioidea line, oothecae have keels. Respiratory structures in the keels, denticles (serrations), and calcium oxalate crystals (which contribute to the rigidity of the ootheca), arose within the polyphagids. In some Ectobiidae there is a marked decrease in the size of the denticles and a tendency for the thickened margin of the keel to lie flat against the micropylar ends of the

This ootheca just starting to form (*Lucihormetica verrucosa*)

This ootheca has been fully formed and flipped sideways.

Ootheca old and new (*Deropeltis paulinoi*)

eggs. The oothecae of the Blattidae and most Ectobiidae possess large amounts of calcium oxalate crystals. However, species of *Blattella* (and other species that carry their oothecae externally during gestation) show a gradual loss of calcium oxalate. All of the Blaberidae incubate their eggs internally, the keel as it occurs in the Ectobiidae has been lost, although in some blaberids greatly reduced keels occur as relics. The oothecae of blaberids are flexible, lack calcium oxalate crystals (rarely the crystals are obscure) and are thin compared to most oviparous species. The greatest reduction of the oothecal membrane occurs in the viviparous species *Diploptera punctata*. Only one group of Australian ovoviviparous species (several genera in the Blaberidae: Geoscapheinae) do not form an ootheca and the eggs are deposited as a jumbled mass in the female's atrium or vestibule.

Although a considerable amount of research on cockroach reproduction has been done, much is still unknown and is worthy of investigation. This is particularly true of species like *Perisphaerus*, where the young cling to the abdominal sternites of the mother for at least two instars. Also, species of *Phoraspis* from Brazil have dome-shaped tegmina, similar to species of *Thorax* and *Phlebonotus*, suggesting that the young nymphs may stay with the mother. Very little, or nothing is known of the biology of the early nymphs of these species.

Relatively little attention was given to the significance of reproduction in cockroaches until McKittrick's (1964) classic study. Her work greatly revised the taxonomy and phylogeny of cockroaches, but some of her ideas are still being debated. It may be fitting to end this brief report with two quotes:

Dorylaea orini oothecae are glued to pieces of wood and coated with chewed up fibers. In crowded caging this wears off or is incompletely formed, but the coating makes the oothecae nearly invisible under natural conditions.

The ootheca of *Polyphaga* are often heavily sculptured.

Polphagid oothecae are usually buried and never glued to objects. The thick shell protects them.

Fresh oothecae are often pale to white to start. Within hours they will turn dark brown.

"It quite saddens me to think that when I cross the river Styx I may find myself among so many professional biologists condemned to keep on trying to solve problems, and that Pluto, or whoever is in charge down there now, may condemn me to sit forever trying to identify specimens from my own specific and generic diagnoses, while the amateur entomologists, who have not been damned professors, are permitted to roam at will among the fragrant asphodels of the Elysian meadows, netting gorgeous, ghostly butterflies until the end of time." (Wheeler 1939: 142)

"We now must determine where we are at—
Which taxa to keep and which ditch?
What orthopteroid group includes this, and what that—
And which do we switch and with which?"
(Kevan 1977)

CAPTIVE REPRODUCTION

External Oothecae

Formation of a thick-walled egg case (ootheca) is the most common method of reproduction among the Blattodea. While most live-bearing cockroaches usually produce a membranous ootheca, it is not the type of ootheca covered in this chapter. Laying eggs may seem a primitive trait but oothecae allow for survival in relatively dry and hostile habitats.

External oothecae each contain from two to more than fifty cylinder-shaped eggs (most commonly between twelve and twenty-four). The largest known ootheca is made by the largest winged cockroach, *Megaloblatta blaberoides*. Its ootheca is approximately two inches (50 mm) in length (scaled photo in Schal et al. 1984). The smallest oothecae are two to three millimeters across and are not necessarily made by the smallest species. Adult *Paratemnopteryx couloniana* are many times the mass of small ectobiids but produce tiny oothecae.

Each egg inside the ootheca has a thin membranous shell and cannot survive exposure to the environment. The surrounding shell of the ootheca is hard and waxy to prevent desiccation and provide protection from external hazards. Because the well-protected eggs may survive even the dryness of a home, the four most successful pests are oviparous. The outer shell, however, does not protect the eggs from stagnant high moisture levels that will kill the eggs inside. The case is restrictive but permeable to air and water.

Females generate a new ootheca every five to twenty one days. The eggs are laid upright at the end of the abdomen and the female secretes a protective coating over them. The formation of one complete ootheca can take twenty minutes to a few hours. Across the top of the ootheca a zipper mechanism (the keel) is created as the ootheca is formed. The keel contains a weak seam that pops open when the nymphs are ready to squeeze out into the world. The nymphs must swallow air in unison to burst this seal. Under circumstances of extreme stress after shipping, lack of food, or lack of moisture, oothecae are still formed. Resulting egg cases are tiny and may contain only a few eggs (or none) or have damaged walls. Even minor damage or deformity of the casing usually dooms most or every egg, especially if the keel is damaged in a way that locks it open or closed.

Hardened, fully formed oothecae are glued carefully into crevices or under bark, placed in a suitable spot, or buried in the ground. In captivity oothecae that are normally glued may simply be dropped on the substrate if sites for egg laying are inadequate. Without moisture or added protection, even within the thick walls of a protective ootheca, eggs of most species dry up. This is the case even if ambient humidity exceeds 40%. A successful way to safely incubate difficult-to-hatch oothecae is to place pieces of decaying wood in the cage. Old wood makes an excellent incubator because it retains moisture and shields the eggs from desiccation and can be placed where there is plenty of fresh air. Females will carefully glue oothecae to the wood and the wood is misted a few times a week. Moisture needs in temperate areas can vary widely by season. In the winter when the dew point is very low, structural humidity can drop below 10% and twice daily misting may not be enough. In late spring and early fall with high humidity and lower temperatures, misting more than once a month might be enough to cause eggs to become moldy.

Oothecae which are normally glued to wood or pasted with substrate are not always properly laid in captivity (usually due to lack of adequate materials or crowding). Abandoned oothecae can be kept in a partly screened container on a layer of moist coco fiber, potting soil, sand, peat, etc. and misted every day or two. Laboratory incubators and homemade incubation chambers are usually unsuccessful with more difficult species but can be very useful to try

a) Burst hatching from an ootheca: this Ectobiidae sp. "little penquin" ootheca was removed from the mother and began to hatch within minutes. (The dark, hardened ootheca is from *Paratemnopteryx couloniana*.)

b) Burst hatching from ootheca: like mantids, blattid hatchlings emerge with a sausage-like shape and have an organ on the head that expands to help them push their way out of the ootheca.

c) Burst hatching from ootheca: now free, they begin to take on the final body shape.

rearing easy species. High moisture and lack of airflow causes sensitive oothecae to become moldy and die (too little moisture and they dry out, anyway). Many egg layers, including all members of the desert cockroach family, do not glue oothecae but deposit them in moist soil. These are usually best kept in the colony enclosure with the dirt kept damp but never soggy.

Oothecae hatch after three weeks to six months depending mostly on species, but temperature plays a role. Many types of tropical oothecae do not hatch if kept below 72° F (22° C) and develop most rapidly at 79-85° F (26-29° C). High heat is not always safer or better since eggs may be easily cooked or dried out. Hatching seems to take forever when starting a colony but, because large quantities of oothecae are produced, as soon as they begin to hatch there is a constant flow of new nymphs.

All but one of the cockroach families reproduce, with rare exception, exclusively through oviparity. Members of the families Ectobiidae, Blattidae, Polyphagidae, Nocticolidae, Lamproblattidae, Tryonicidae, and Cryptocercidae produce thick-walled external oothecae. Three of these families are commonly maintained and their oothecae can often be identified to family by the structure or the way they are deposited.

The Ectobiidae is the most diverse cockroach family and ootheca are somewhat variable. Most species form an egg case carried at the end of the female's abdomen until a suitable spot is found to place the eggs, but some carry the ootheca partly retracted within the abdomen (very rarely fully retracted). The Ectobiidae includes the pale-bordered field cockroach (*Pseudomops septentrionalis*), the banded wood roach (*Parcoblatta zebra*), and (the longest of all roaches at just over 4 inches, 102 mm+) *Megaloblatta longipennis*. Although oothecae sometimes resemble the sculptured egg cases of polyphagids, ectobiid oothecae lack large teeth on the keel and are nearly always curved (curved on the opposite edge of the keel) unless they are very short. This family includes most of the rare exceptions to the brown-colored ootheca—white or gray oothecae (*Balta* and *Ellipsidion*) are glued to branches and green *Supella* ootheca are attached to anything. When glued, they are pasted along the bottom (opposite side of the keel) to surfaces. Ectobiid ootheae are usually the quickest to hatch, often within three weeks.

Eurycotis lixa ootheca

Australian cockroach with ootheca
(CC) Cyron Ray Macey

The Blattidae is the third largest family but the egg cases are hardly variable by species or genus. Blattids are usually large species (>1″ or 2.5 cm) that form a globular, massive ootheca with a thick, often lightly dentate, keel. The sides are smooth and the color is dark brown, nearly black. The harlequin cockroach (*Neostylopyga rhombifolia*), zebra cockroach (*Eurycotis floridana*) and the beautiful African *Deropeltis* all form very similar-looking oothecae. Females glue each ootheca sideways to a surface and then chew up materials that they paste over the exposed side of the ootheca. Some females paste ootheae into the substrate. This brief maternal care of the ootheca is not always seen in captivity as it depends on available surfaces and materials. Gestation is usually two to three months.

The Polyphagidae comprise less than half as many species as the Blattidae but the oothecae are many times more variable. The egg case structure can identify genera or even species. Oothecae of the elegant Burmese beetle mimic (*Ergaula capucina*), spectacular domino roaches (*Therea* spp.), and the strangely hairy desert cockroach *Eremoblatta subdiaphana* are easy to tell apart. Oothecae produced by this family are usually intricately shaped with ridged outer walls, strongly flattened or toothed keels, and a flange at the end which is used by the female to carry it (held by the rear abdominal appendages). Females carry around the ootheca at the end of the abdomen for a day or two as they look for a place to bury or deposit it. Deposition is primarily based on moisture since females carry oothecae around sometimes for weeks if the enclosure is very dry. Ootheca are not glued to objects. These species usually require the longest gestation, often three to six months.

The tiny Family Cryptocercidae has gone from three to maybe a dozen species in the last decade but they are all members of a single genus (*Cryptocercus*). The secretive Cryptocercidae form small oothecae and glue them into their tunnels where they care for them.

Live-bearing Roaches

The second most common method of reproduction is called ovoviviparity or live bearing. Females begin with the external production of a large, membranous egg case (except for *Macropanesthia* and other Australian Geoscapheinae that transfer eggs directly from the ovaries to the uterus). The egg case is formed with the eggs in vertical rows, usually in twos, but often jumbled and imperfect. Once fully formed, the structure is then rotated sideways and drawn back into the abdomen for incubation within a moist uterus. The outer casing is more of a membrane than a theca and the eggs dry up and die rapidly outside the protective abdomen of the mother. Live-bearer oothecae possess no color, no keel, and no hardened outer shell, and the casing is often incomplete (it may cover only some of the eggs). Since the eggs temporarily leave the body, this method is sometimes called false live bearing (versus true live bearing in Geoscapheinae).

Healthy oothecae of live-bearing species are not always in rows of two (*Nauphoeta cinerea*).

Thirty minutes to a few hours is the time required between the production of the egg case and drawing it back into the abdomen. The eggs then remain inside the female for a few weeks to as long as nine months. Hatching usually takes less than an hour and immediately follows evacuation of the eggs. If the eggs protrude from the female's abdomen other than during the initial formation and retraction or during hatching, the eggs are being aborted. Sometimes an improperly formed ootheca will stick slightly from the abdomen and only the eggs near the opening dry out and die. Damage to the uterus can prevent a female from reproducing at all. Disturbances during ootheca formation may damage the structure or prevent it from being retracted. Poor nutrition,

Aborted eggs from a poorly formed *Gromphadorhina* ootheca

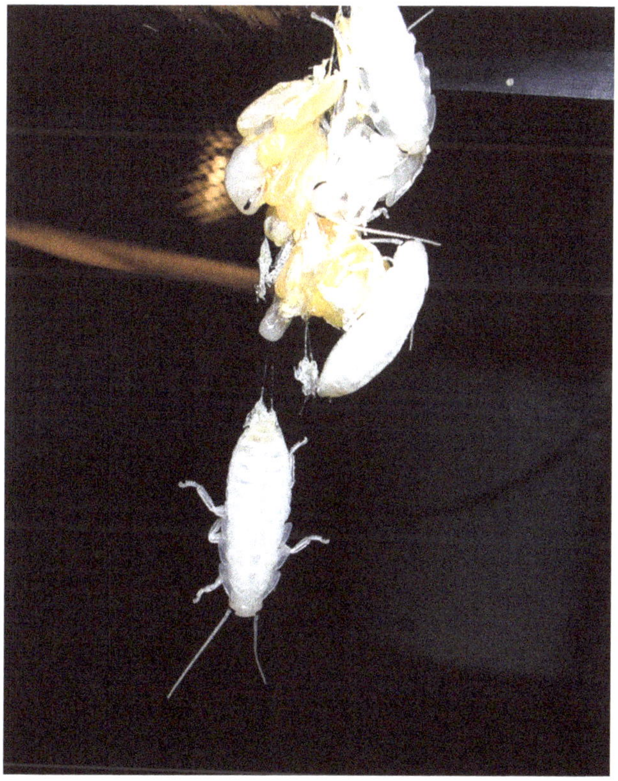

Hisser babies busting out, as the mother is hanging upside-down from a screen lid; this is very unusual, but all the developed young emerge successfully.

overcrowding, low temperatures, and lack of food or water cause abortions. It is possible to watch an aborted ootheca dry up and start to wrinkle, sometimes within minutes. Attempts at salvaging aborted oothecae are never successful unless it was starting to hatch. When placed in humid incubators they just grow moldy (even if mold inhibitor is added).

Live-bearing females often abort eggs when being eaten by a predator. Under extremes stress the ootheca is aborted and hatches prematurely (unless underdeveloped). The babies can escape as mom is being eaten. A similar response is seen in egg-carrying species. If the eggs are not old enough the ootheca is still aborted but nothing happens except the eggs dry up and die.

The live-bearing strategy is not as perfect or advanced in the ways it is often suggested. Live bearing is commonly said to produce fewer, better-protected eggs, but this is not true because live bearers often live longer or have more eggs per ootheca. Most of our live-bearing species produce just as many offspring as the egg layers though usually over a longer period. Some live bearers can exceed the total production of many egg layers in a much shorter period. A female *Gyna* can bear more

Young blaberids like these hissers often feed on afterbirth provided by their mother.

than 200 in a single brood, close to the lifetime production of many egg layers. Live-bearing species do invest a lot of energy and often time on each brood. They must carry the growing weight of the eggs around inside of them until hatching, which can be as long as nine months or as quickly as three weeks. While carrying developing eggs, females cannot produce more oothecae. The eggs often take up a larger relative portion of the adult roach's weight and take up moisture from the mother as they grow. Females need to build up extra energy to produce, transport, and gestate the eggs. Under slightly adverse conditions females may hold oothecae indefinitely, which can be frustrating.

Although successful and varied in the tropics, live bearers are rare in deserts and temperate areas. Live bearers hardly even make the pest list because the strongest are decimated by constant low humidity. Almost every native North American and European cockroach is an egg layer. This does not seem to be directly related to the inability to create independent weather-protected oothecae. All cold temperate species I know of do not survive the winter within protective oothecae like mantids, after the adults have died, but rather overwinter as nymphs.

Live-bearing species are able to help protect their offspring in ways the egg layers cannot. First, eggs are far safer inside a female since they are protected against predators, parasites, and the harsh external environment. External oothecae are commonly attacked by wasps in the wild, while a wasp that attacks eggs within a live bearer has yet to be discovered. An ootheca cannot move itself out of the

Motherly care in *Archimandrita tessellata*

hot sun or drowning downpour of rain. Nor can an ootheca hiss, emit defensive secretions, or run away from a hungry bug. Second, a female can find ways to aid her young when they first hatch if she happens to be nearby every time. In most species the female guards her babies for at least the first few minutes of their lives as the hatchling exoskeletons harden. In many species a protein rich secretion is produced from the tip of the female's abdomen immediately following the emergence of young. This 'afterbirth' is hungrily taken as an easy meal by the teneral nymphs. Species of *Aptera*, *Macropanesthia*, *Panesthia*, *Rhyparobia*, and many other genera take care of their young for weeks, months, or even years. (The peculiar Cryptocercidae are the only egg-laying species known to take care of their young and they only produce one or two broods in a lifetime.)

Corydidarum females shield the pale newborns with their hardened body shells.

The Family Blaberidae contains nearly all the ovoviviparous cockroaches. The Blaberidae is the second largest cockroach family with

Corydidarum females still protect nymphs a month later (they do not seem to differentiate between their own or other females` young).

Large nymphs continue to cluster with adults for protection.

Simandoa conserfarium adults and nymphs

Macropanesthia rhinoceros, 4 days old with small eyes and long antennae (May 2, 2016)

more than a quarter of known species. Oothecae are widely variable in size and structure only because the egg size and numbers vary. The outer surface is membranous, usually see-through, and sometimes incomplete. There are commonly two dozen to four dozen young per brood, but there can be hundreds. Live bearers include common large pets like the various Madagascan hissing cockroaches and common feeders such as *Blaptica dubia* and *Panchlora nivea*. The heaviest of all, the rhinoceros cockroach (*Macropanesthia rhinoceros*) from the subfamily Geoscapheinae is also live bearing, but members of its group do not form oothecae at all. Beyond the Blaberidae, just a couple of oddball members of a handful of tiny genera of Ectobiidae, including the Australian *Pseudobalta* and African *Stayella,* are live bearing.

The only known viviparous species of cockroach is a member of the family Blaberidae. The Pacific beetle mimic cockroach (*Diploptera punctata*) is viviparous and produces a small number of eggs that must get nutrition as well as moisture from the mother. Underdeveloped eggs are nourished within the female's body until growing into fully developed nymphs. Over the course of three to four broods it is unusual for a female to produce more than fifty offspring in her total lifetime.

PARTHENOGENESIS

Reproduction without males is normal for certain orthopteroids, namely the phasmids of whom nearly every one of the more than 3,000 species can reproduce this way if males are unavailable. In the case of Dictyoptera this mode is rare but standard for a few species and facultatively possible (they are not known to reproduce this way in the wild but in captivity eggs may develop if females are unmated). However, most dictyopteran species cannot reproduce without male fertilization and of the few that can, many cannot continue on for long since the survival rate is low. Bell and Roth (1956) showed it was possible for some of the pest species like *Periplaneta americana,* but possible does not mean realistic or repeatable. Their report begins by saying other researchers have considered parthenogenesis not possible in many of these species because ". . . hatching of eggs was used as a criterion of parthenogenesis." I have isolated female *P. americana* 'white eye' on numerous occassions and have never seen an egg hatch from an unmated female. *Blatta orientalis*, on the other hand, can form large colonies without males. *Pycnoscelus surinamensis* is a familiar and widespread parthenogenetic species whose reproductive methods have long been familiar (Chopard 1938). Still, males are seen in some Australian populations (Rentz 2014) and *P. surinamensis* in Hawaii have males (Roth and Willis 1960). *Pycnoscelus nigra* likewise is normally parthenogenetic but other members of the genus cannot reproduce this way (i.e. *P. femapterus* and *P. striata*). *Polyphaga obscura* and *Polyphaga saussurei* colonies have been without males for maybe twenty years but in natural populations males are present.

INBREEDING

Inbreeding is not usually a topic the blatticulturist has to seriously consider. All of the old stocks have been inbred for decades and if anything they have become easier to maintain. Of course it is important to remove deformities and defects whenever possible to strengthen stocks. The most practical concern related to the concept of inbreeding is minimum colony size. If only one or two of each gender are kept each generation, at some point the stock will certainly die out since not every adult animal is fertile, functional, or healthy. Too small of a

These female *Pycnoscelus nigra* nymphs are not only sisters, but also clones of their mother.

Eurycotis decipiens with segment deformity (2014)

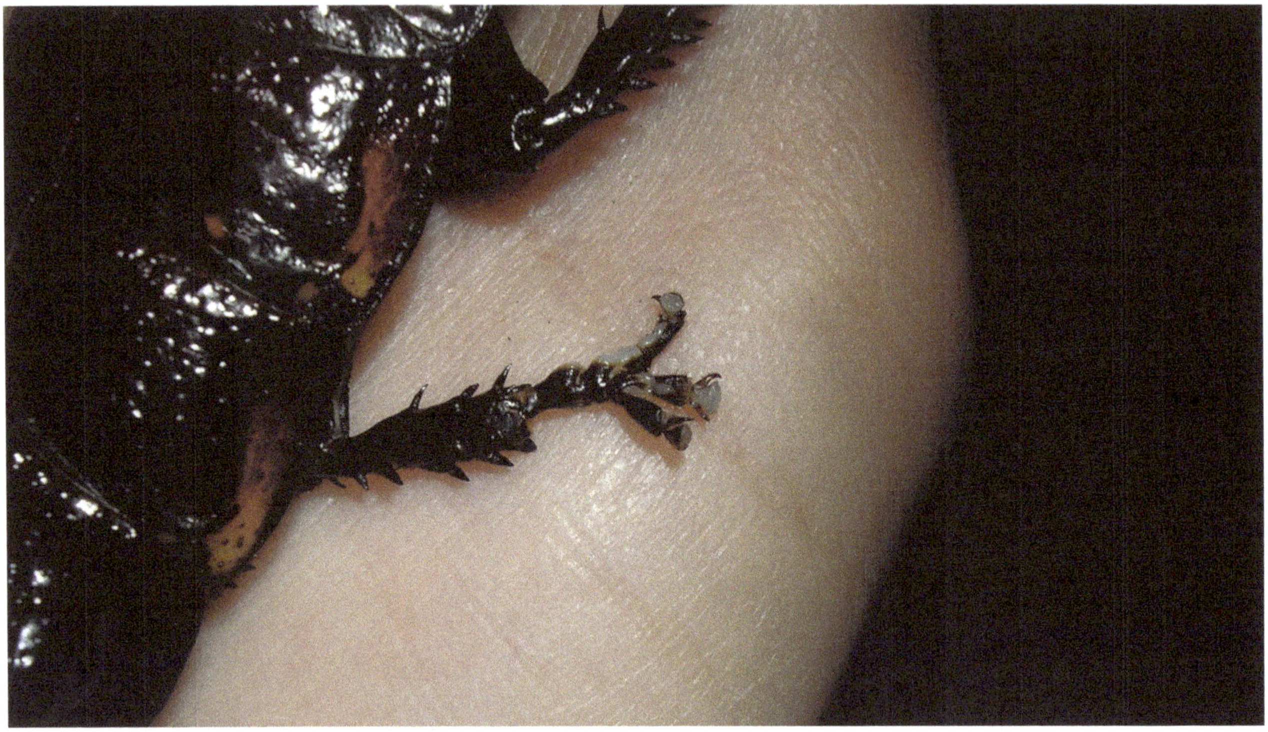

Deformed tarsus, three copies (2009)

Therea petiveriana with various expanded white patterns (2015)

population prevents continued survival if a few are accidentally smashed in the lid, dried out, etc. A few dozen individuals is usually the minimum any culture should maintain. I wish I could blame a single one of the beautiful or amazing species I have lost on the accursed myth that inbreeding is detrimental to their health. Most of the time animals are lost due to neglect such as forgetting to water for too long, but foolish mistakes take their toll as well (like trading off too many *Deropeltis* or V-horns or keeping *Macropanesthia* nymphs in deep substrate before understanding their requirements).

Common segment deformity can be seen on a tiger hisser (2008). Similar segment formation problems are seen in beetle larvae, pseudoscorpions, millipedes, and other invertebrates.

ABERRATIONS, VARIATIONS, AND COLOR FORMS

Breeding for size, color, or other features can be a rewarding aspect of keeping roaches. The easiest to select for is color and there are a few established color cultivars of long-term captive species. The most familiar and stable are the black version of *Gromphadorhina portentosa* and the yellow form of *Gyna lurida*. These may be attributed to recessive genes because once the standard coloration has been completely removed, the offspring are 100% black or yellow. *Blaptica dubia* have been bred to produce a pale, yellow-orange variety, but the only known pied *B. dubia* died for uncertain reasons. Laboratories have produced a shiny, deep black cultivar of the pestiferous *Blattella germanica*. I have selected *Therea petiveriana* for excessive white coloration over a few generations with good success. Interesting forms may slip through the cracks: only one specimen of *Panchlora nivea* with a black pronotum has ever been seen (in 2008) and it was accidentally killed.

Opisthoplatia genetic segmentation problem (2008)

Aberrant eye color is usually a recessive gene and is easily isolated when encountered

Lavender eye mutation in *Blaberus craniifer*

Yellow *Gyna lurida* (2002)

Deformed partial albino *Blaptica dubia*.
© Ryan Nefcy

A unique *Panchlora nivea* female with a black pronotum (2008)

(McMonigle and Willis 2000). So far, only a single colored-eye form is widely kept, the 'White-eye' *P. americana*. Lavender eye forms of *Blaberus craniifer* and *P. americana* have also been documented (McMonigle 2015d). Another strange deformity is the 'prowing' form of *Blattella germanica* isolated by Roth in 1964 and still maintained in laboratories (Tanaka and Ito 1997). The prowing is a variable, non-recessive, miniature wing-like deformation of the pronotum. Well-developed prowings are reminiscent of the prehistoric paleodictyopteran *Stenodictya* which had a third pair of small wings on the pronotum.

Size has proven to be the most difficult feature to select for in captivity. Massive hissers have been produced from various stocks but the factors involved may be more influenced by male dominance, sociality, and caging. Not every stock will throw large animals, but the largest stocks only throw a few super-massive males per hundred. I have culled small males from my hisser stocks for decades and have yet to see a return on my investment. However, as luck would have it, if the largest males are pulled out over a few generations, they seem to disappear altogether.

A gyandromorph (half male, half female specimen) is seen on rare occasion (2001?). This genetic anomaly cannot reproduce.

Therea olegrandjeani with expanded white markings (2012)

Therea petiveriana with extra white (2009)

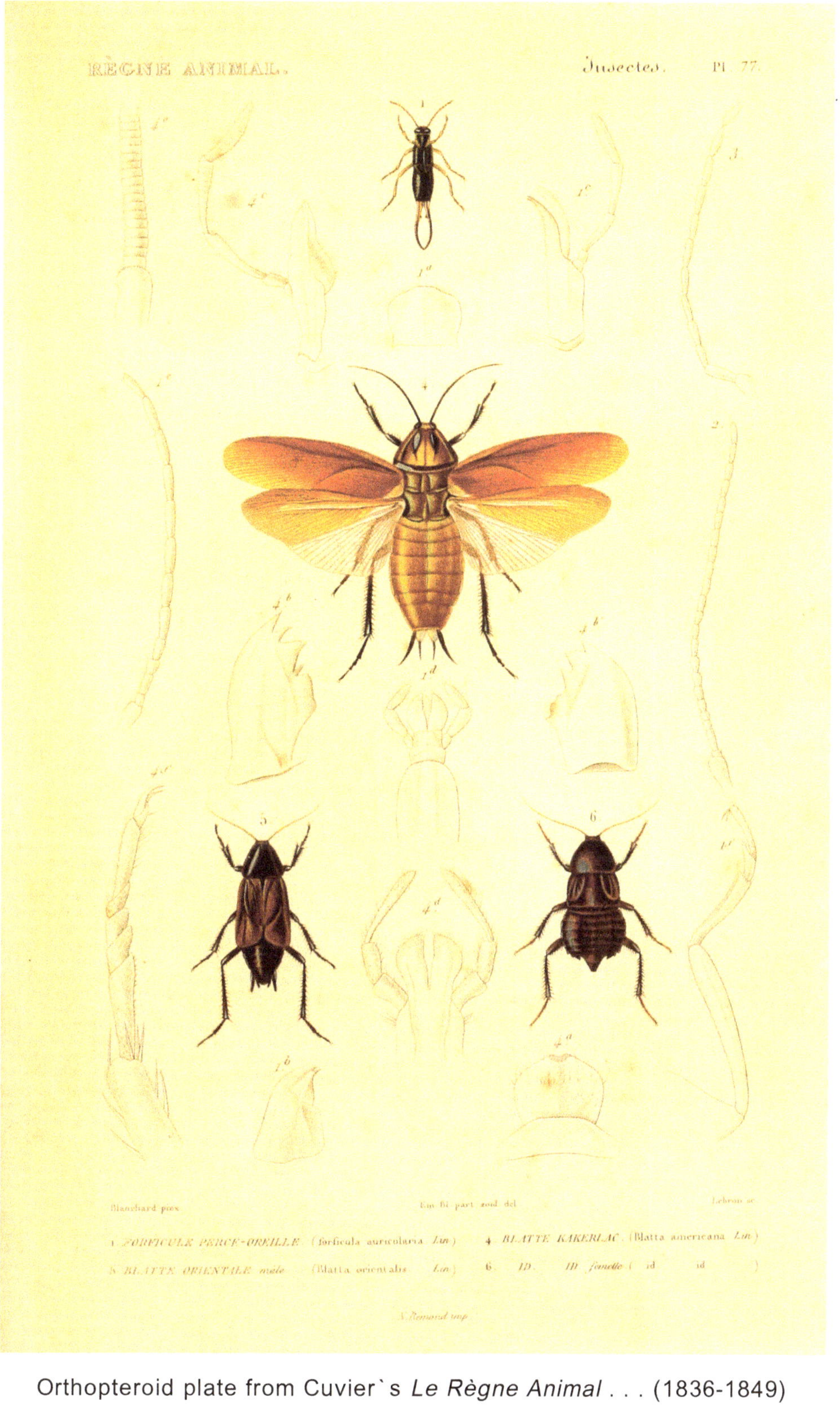

Orthopteroid plate from Cuvier`s *Le Règne Animal* . . . (1836-1849)

Ellipsidion nymph, New South Wales, Australia (CC-BY) John Tann

Unidentified roach nymph, Queensland, Australia. (CC-BY) James Niland

HUSBANDRY

飼養

ENCLOSURES AND ARTIFICIAL HABITATS

Keeping animals contained is the most important aspect of an enclosure. The greatest worry for the ordinary enthusiast is that escapees of the various pet species will die or become fatally damaged after only a few days loose in the home. All the work raising them, the enjoyment of watching them, money invested in their purchase, and time spent caring for them would be wasted. Some specimens are very expensive, while others can be irreplaceable. Although species in many genera, including *Blaberus, Gromphadorhina, Byrsotria*, etc., do not endure long in a home environment, if they are lucky enough to escape in a lush greenhouse or planted lizard cage, may survive quite well. Modern structures are rather hostile and dry environments for species adapted to forests and caves. If the environment does not dry them out immediately, the pet cat may play with them to death or they can be accidentally stepped on. All the major structural pest species are egg layers, though most egg-laying species dry out and die more rapidly in a home (sometimes within hours) than the relatively bulky live bearers. One live-bearing species, *Pycnoscelus surinamensis*, can infest homes on rare occasion. This creature reproduces indoors within the soil of houseplants, but dries up and dies when the potted plants are removed. Certainly a worry for some hobbyists is the possibility of pests (not pets) escaping and setting up permanent residence. Without correct caging, this would be likely to happen to those more zealous enthusiasts who keep the small number of structural pest species like the American or German cockroaches. Even the unique cultivars like American white eyes and black Germans probably pose the same risk as the standard variety.

When it comes to escape, all roaches are not created equal. Members of the genera *Blaptica*,

Hisser hut constructed of Masonite, held together and spaced with a 1/4" threaded rod and nuts on each corner

Large numbers of *D. paulinoi* can do well in a very small area.

Many species molt in proximity without great risk from their cage mates (*S. conserfariam*).

Therea regularis in a glass terrarium

Panesthia, Blaberus, Blatta, and many others cannot climb smooth surfaces. A container made of plastic or glass easily contains them, even without the use of a lid. Members of the genera *Parcoblatta, Blattella, Eurycotis*, et al., are able to climb smooth glass, plastic, and metal. Some genera including *Panchlora* and *Gyna* have nymphs that cannot climb, while adults can scale anything. Only the adult males can climb smooth surfaces when it comes to most desert-living polyphagids. The majority of species have large sticky pads (euplantulae/pulvilli) on the underside of the basal tarsal segments, but climbing species also have a large pad (arolium) between the claws. For all roaches a lid is suggested, because even those that lack glass-climbing ability still climb out if a piece of egg carton or wood is accidentally placed against the side, close to the top.

Nearly all species capable of climbing smooth surfaces can be kept in check by the use of petroleum jelly, silicone gel, or other lubricant barrier. A thin layer of the lubricant is applied to the top inch or two around the sides of the cage. Each attempt to climb wipes a small amount of lubricant away. The barrier may need to be reapplied only once or twice a year. Cages containing huge cultures or especially nimble climbers may require reapplication of the barrier every week. Dust, flies, frass, and moths get stuck in the barrier, allowing circumvention. Although a slippery barrier works to successfully contain most species, a few types can climb up oily materials (mostly just first instar nymphs). Small specimens of certain species get stuck in petroleum jelly and die. Various greases and vegetable oils have been tried with varying successes.

Liquid Teflon is an alternative barrier that is less messy and is not sticky. It lasts much longer since it is not easily wiped away. Some species can walk up a liquid Teflon barrier, but others are better controlled. Multiple attempts at escape do not wipe it off the surface and dust or dirt does not stick to it if the cage is tilted. Liquid Teflon is more difficult to acquire, vastly more expensive, and can be unsightly on clear container walls (the dried film is opaque white).

Liquid Teflon painted on cage walls to prevent climbing is unsightly on clear caging (also notice the aborted *Corydidarum* ootheca)

Even though affected by a barrier, some species require more than a barrier for confinement. Hatchling nymphs of *Eurycotis* species are able to easily walk over lubricants (first

Bucket cages and others

instar hissers and *Corydidarum* may get past it). Adult *Gyna* spp., *Panchlora nivea,* and *Schultesia lampyridiformis* are able to fly. Even petroleum or silicone jelly does not keep them captive as they can fly to the lid and climb out through spaces between the lid and container. A bucket cage keeps even these pesky escape artists confined, without chance of escape.

Bucket cages are excellent cages for roaches. They are escape-proof as long as they are made with the particular species in mind. These enclosures are made from a bucket (plastic boxes rarely seal well), a fitting lid, and some window screen fastened with a hot glue gun. Tiny buckets from fish food containers or deli-cups can be used for small species. Bucket cages are easily stacked (both empty and full) and can be kept almost anywhere. Moreover, they have the benefit of being inexpensive.

To Make a Bucket Cage

1. Find a bucket with an airtight lid.
2. Cut a square or hole in the lid and on one or more sides of the bucket. Holes are usually two or more inches (>5 cm) square. Vent size will need to be estimated based on chosen vent material, ambient humidity, airflow, and species.
3. Measure and cut pieces of screen that are one-quarter to one-half inches larger than the holes.
4. Use a hot glue gun to carefully attach screen over holes.

Some bucket cages from the 1990s

Important Points

Use thick, sturdy buckets because thin buckets tend to crack over time.

Hot glue sticks to most plastics, but not all. A dot of glue can be tried on the plastic before cutting holes. Other adhesives like epoxy or silicone can be used, but they are messier, more difficult, more expensive, take a long time to cure, and do not necessarily work as well.

Cloth "micro" screening glued over hole in a 32 oz deli-cup provides ventilation and prevents escape of very tiny nymphs.

Mixed hardwood leaf litter is commonly placed with burrowing species, especially polyphagids and *Gyna*.

Consider the integrity of the container when placing vents. Do not place too close to the edges or lid. Do not make the vents so large the container collapses from minimal stress. Even minor collapse can warp and separate the glue over time.

Glue screen from both the inside and outside.

For smaller species, microscreen or inorganic cloth such as nylon or polyester should be used in place of window screen. If cloth or plastic material is chosen, it is helpful to glue a layer of regular mesh metal screen on the inside or outside. Common window screen mesh is 18 x 14 wires per inch and allows for good ventilation, while microscreen mesh (like that used for coffee filters) can be 100 x 120.

The "Australian balled" roach was an unidentified species kept in the mid 1990s that quickly ate through any nylon or cloth venting material. It was the size of *Periplaneta*, but adults had no wings and the end of the abdomen was always covered in a sticky coating. This tiny photo may be the only trace that this stock ever existed.

Metal screen is suggested, although it is more expensive and difficult to acquire in fine mesh. Some blattodeans chew through nylon screen and muslin cloth. (The worst culprits are rarely seen in culture: members of the subfamily Panesthiinae and the Australian 'balled roach'). Standard mesh, aluminum window screen can be used as a barrier to protect muslin cloth or screening, though small nymphs of some species may get permanently stuck between the screens. Burrowing roaches including *Gyna* and *Pycnoscelus* do not normally chew on soft screen materials. However, they will chew holes in the screen if the ports are partly buried below the substrate.

Metal screen must be used if a 60-watt light bulb is placed over the cage. Plastic screen will melt and cloth screen may catch fire. The metal screen-covered hole in the lid should be as large as the reflector. The bulb must be far from plastic and other flammable materials to prevent melting or fire. See manufacturer's recommendations for any bulb used.

A petroleum jelly barrier is still useful in keeping glass climbers from escaping when the lid is removed for feeding or cleaning.

Many of these enclosures contain cockroaches

A variety of enclosures have been used to house cockroaches. Large plastic totes are inexpensive compared to glass terrariums and are commonly used to house large, non-climbing feeder species by those who have little interest in looking at their animals. Plastic jars and cups may be adapted like the bucket cages and are primarily used for smaller species. Glass jars are hardly useful but glass terrariums (sold as aquariums) make nice cages.

All-glass terrariums (sold as aquariums) have long been popular for large species because it is easy to design an aesthetically pleasing enclosure. I have seen a few handsome living room displays in 55-gallon (208 L) terrariums. A *Blaberus giganteus* colony can become a fantastic display inside a 30-gallon (114 L) high terrarium with a screen lid and lichen-covered cork bark decorations. Glass terrariums (like most plastic boxes) are not escape proof, because the available lids do not seal well. Species like *Panchlora nivea* can be contained by adapting a metal window-screen lid (small mesh lids are difficult to find lately, with most available lids now constructed from a much wider, heavier mesh), by lining the underside edges with Velcro tape. The opposite side of the Velcro tape is then placed along the top edge of the plastic aquarium frame to provide a tight seal. Care must be taken to leave no gaps. A Velcro seal can also be used to contain hatchling hissers.

Most roaches benefit from an inch or two of substrate on the bottom of the enclosure. Possible substrates include potting soil, wood mulch, dead leaves, vermiculite, shredded coconut shells, wood chips, and gravel. Most substrate materials provide a buffer against low moisture levels, which are deadly for a number of tiny egg-laying species, oothecae, members of the blaberid subfamily Panesthiinae, cryptocerids, etc. Furthermore, most species chew on whatever substrate is present when food is in short supply. Rotten wood substrate or a layer of dead leaves provides extra nutrition for the roaches and holds moisture efficiently but may grow mushrooms and mold (usually harmless) if kept too damp. Shredded coconut shell changes from dark brown to light reddish-tan as it dries. It provides limited nutrition and rarely grows significant mold. Vermiculite, gravel and wood chips provide no nutrition. The keeper should pay attention to the blattodean reactions to previously untested substrates. Some materials, such as redwood chips or shavings, irritate numerous species and send them up the sides of the cage. Cedar and redwood are toxic for most animals.

Unlike rodents and most vertebrate pets whose cages stink like mad after a few days without cleaning, roach cages often do not smell too bad after many years without cleaning. Roaches with strong defensive smells, like the Florida skunk roach, do stink terribly, but only when seriously disturbed. The smell dissipates after a few minutes to a few hours. If the tank smells badly, the odor is more likely due to old rotting food than to the roaches. However, rotting cockroach corpses smell horrible except under the driest of conditions. A few species impart strong odors to their enclosures all the time, specifically the American cockroach and other *Periplaneta*. Species odors are not always terrible. *Gyna capucina* cages emit a very pleasant, sugary odor.

The cleaning schedule for caging depends on the size of the cage, number of individuals, species being kept, and food provided. Hissing roaches tend to build up a thick bed of frass after a few years, especially if fed leafy greens or leaves that are eliminated primarily as indigestible cellulose pellets. Many species keep their own cages clean, but *Blaberus* usually will not eat the shed exuviae of late instar nymphs and many do not eat their dead. *Diploptera punctata* and *Phoetalia pallida* are among the worst, since it becomes difficult to pick the living animals out from the cages after a year if dead bodies are not removed. The living are easily lost among the layers of dead ones. One *Panchlora nivea* enclosure (a 10-gallon or 38-liter glass terrarium) has produced tens of thousands of individuals over twenty-one years and has never been cleaned. It remains functional and does not look a lot different than it did a few months after it was set up.

Keep in mind the requirements, behaviors, and abilities of each species when designing the enclosure. The Madagascar hissing cockroach (*G. portentosa*) is often kept in less than adequate cages by new hobbyists. This species usually restricts cage-climbing antics till after the lights go out and the unaware hobbyist is fast asleep. Adults are easy to contain with most screen or vented lids and a Vaseline layer to prevent climbing. Early instar *G. portentosa* nymphs are escape artists that can easily squeeze through gaps under screen lids or plastic vents. Some stocks of common feeder species like *Nauphoeta cinerea* and *Schultesia lampyridiformis* have been trained over years to avoid climbing (those that climb are immediately fed off, generation after generation, until the behavior changes) but they never lose the ability.

TERRARIUM DECORATIONS

Various hides can be purchased, built, or collected. Paper egg carton sheets are the most common decor, commonly available for those culturing feeder crickets. Electronics packaging often has molded cardboard sections that are similarly useful and come in a wide range of shapes and sizes. Toilet paper and paper towel rolls are too smooth to be of much value. Pieces of decayed wood (not too soft) are often used as decorations for small species that do not burrow, since they retain moisture better than paper products. Furthermore, females of many egg-laying species protect oothecae by pasting them to depressions in the wood. Terrarium decorations readily available from local pet stores (cactus skeletons or cork bark pieces riddled with holes) can be selected to shelter small nymphs, increase areas for laying, and protect oothecae. Most pieces of tree bark and coconut huts are too smooth to be useful for tiny nymphs or egg laying, but they make excellent hides for burrowing genera like *Archimandrita* and *Lucihormetica* to pile up below. Pine is usually warned against for use in invertebrate enclosures, but I have used the same partially decomposed white pine logs in caging for *Blaberus, Eublaberus*, and *Panchlora* for a few decades because they are not

Wooden decorations increase the available surface area in a bucket cage.

Wooden decorations in a glass aquarium

readily chewed apart. Roach huts, built from sheets of Masonite with nut and bolt spacers in-between, are a great way to expand the usable surface area within a given enclosure. These are especially useful for large cockroaches that prefer not to burrow in substrate, such as *Gromphadorhina* and *Opisthoplatia*.

Most species will chew up cork bark decorations over time.

Decorations serve an additional purpose beyond providing hides. Winged and non-burrowing roaches should be given a choice of molting surfaces such as large upright logs or vertical pieces of paper egg carton. The surface should be rough and easily scaled by the intended species. Paper only works with very small species. Angled surfaces that tend toward perpendicular should be avoided because wings can fold unnaturally backwards. Small branches and sticks are generally too small, too round, and too smooth to make good molting surfaces for anything but very tiny species. Related insects like praying mantids, grasshoppers, and phasmids are prone to bad molts that result in deformities, missing appendages, and death. Nymphs, wingless, and short-winged roaches seldom experience similar difficulties during molting because they do not need to hang upside-down (for an important exception see *Macropanesthia* in species list). Nevertheless, without adequate molting surfaces, long-winged roaches, notably *B. giganteus* and *A. tessellata,* are very likely to emerge from the final molt with deformed wings. Even small *S. lampyridiformis* and *P. nivea* need adequate surfaces for the wings to inflate properly. Wings disfigured during molting can bleed profusely. The damage and loss of blood can sometimes result in death.

Lack of adequate molting surfaces is the primary origin of molting problems. Overcrowding can lead to severe deformities, while cage decorations are directly related since they multiply the livable surface area within a given container (substrate depth increases the livable area for burrowing stages or species). Starvation, dehydration, and low temperatures can cause molting problems even if the cage decorations and hides have been chosen with absolute precision. If most adults emerge with deformed wings, and tons of roaches are in a small cage, it is time to split part of the culture. Although wing deformities and bad molts are caused almost entirely by the environment (rather than genetics), culling malformed individuals from a culture often decreases molting problems in future generations (McMonigle and Willis 2000). The reverse is also true.

TEMPERATURE

Temperature can have a major effect on the speed of reproduction, growth, and death. Nearly all species can be kept in a cool basement and will grow reasonably well at 60° F (16° C). However, they grow slowly and may never reproduce. Many species simply will not give rise to young at temperatures below 72° F (22° C). The problem arises in development of the eggs, whether internal or external. Even the oothecae of temperate species will not develop

Dead *D. punctata* all over the cage floor died of old age.

It is difficult to perfectly mimic a natural habitat. This Ectobiidae specimen has found shelter with a salticid in a rainforest in Panay, Philippines. © Christian Schwarz

at much lower temperatures so they spend the winter as nymphs. In cases where oothecae can survive winter, they still do not hatch until temperatures have risen. Internally maintained eggs of most species associated with caves like *Blaberus, Eublaberus*, and *Simandoa* do not develop at low temperatures. Deep-cave dwellers should have the ability to develop near fifty degrees, but this may be why cave dwellers are all but entirely absent in temperate areas.

Under the cage heating pads or heat tape can be used to speed up reproduction and growth, but these systems can be expensive and care should be taken to provide a natural temperature gradient (most roaches move down to avoid heat, which is a problem if the heat source is at the lowest point). Central heating or individual incandescent or halogen light bulbs are preferable. Heat provided by light bulbs is stratified and allows roaches to regulate their own temperature. The majority of species achieve optimal growth when kept between 77-86° F (25-30° C). An entire room can be kept in the mid-eighties, but this is usually only feasible for enthusiasts with already heated reptile or fish rooms. High temperatures are not necessarily good, since they cause food to spoil more rapidly and can lead to rapid and excessive drying of the habitat. Many species are more difficult to maintain at optimum growth levels because closer attention to food and water is required. A few days of neglect can dry out and kill an entire colony. Genera including *Cryptocercus, Macropanesthia*, and *Panesthia* (or their gut fauna) can die from temperatures approaching 90° F (32° C). Raising a low temperature because a colony is not breeding, such as from 68° F (20° C) up to 75° F (24° C) can be very successful, but raising temps from 77° F (25° C) to 86° F (30° C) usually destroys a faltering colony because temperature was not the problem, and the raised temperatures lead to more difficulty maintaining survivable conditions.

A number of cockroaches enjoy sunning themselves, like snakes and lizards do. A light bulb with reflector can be provided directly above the cage for difficult live bearers and favorite species. Gravid females sit on rocks or logs under the light and place their egg-filled abdomen at the brightest, warmest point of light. Although hotter temperatures speed up gestation, growth, and old age, some species appear to live notably longer when provided a light for basking. The adult lifespan, as well as the luster and coloration of the exoskeleton, is improved for *E. decipiens*, *G. portentosa*, and possibly other species. If the blatticulturist provides a varied diet, hiding places, and moisture, and still has difficulty getting young, a basking light may be all that is needed. Basking lights are not always useful. The more arid humidity gradients of the lighting caused an isolated group of *Deropeltis paulinoi* to stop developing over a number of months with limited improvement even after increased watering frequency. A control group of eighteen stayed in the original enclosure, which lacked the basking light. They required less watering, developed normally, and all survived. The experimental group of eighteen was placed in a cage with a basking light. They seemed fine at first, but six months later there were four surviving half-grown nymphs. More heat, more light, more food, more attention, do not always have the desired results.

It is important to place the basking light in a safe place (not too close to flammables) and to choose an appropriate size for the cage. The operational element is heat, not light. The old standard incandescent 60-watt bulbs have been replaced with 43-watt halogen bulbs that are only slightly cooler. LED and CFL bulbs provide little directional heat and are unlikely to raise the temperature even a few degrees from a distance. Ceramic infrared heat emitters last longer than light bulbs and can be left on 24 hours a day without interrupting day and night cycles. These likely work as well as basking lights but they usually come in 100-watt or 150-watt models, which is excessive for a cage smaller than 20 gallons (80 L). It can be more difficult to source 50-watt or smaller emitters.

This bad molt was caused by lack of moisture. This *Blaptica dubia* may live a long time but will never recover.

HUMIDITY AND WATER

Humidity (water vapor) is not the same thing as liquid water—cockroaches cannot drink from the air. It is entirely possible for them to become dehydrated when the ambient humidity is 100%. Still, atmospheric and condensed water are related because extremely low humidity dries out the animals as they breathe. Fluids must be replaced and extreme dryness cannot be countered by availability of moisture alone. Airflow is also an important part of the equation, since even in humid conditions liquid water can evaporate quickly if there is a fan or vent nearby. In nature cockroaches survive dry periods by shielding themselves under bark or burrowing into sand or dirt. Excessive humidity without adequate ventilation is as deadly to most roaches as it is to mantids and grasshoppers. It can take many days or weeks of stagnant humidity to kill all the specimens in a culture container. The risk of death depends on the interplay of temperature, airflow, cage decorations, substrate species, age, size, gender, and moisture content of the food. These creatures (like humans and most living things) can survive and even thrive over a wide range of conditions, so claiming an exact humidity percent even for an identified species, age, and gender is misleading, if not dangerous, when other factors are not also specified. Looking up the natural range can be used as a basic guide, but only if it is specific to the microhabitat of the species. Even then, remember the vast majority of animals that are born in nature die before reaching maturity as a direct result of natural conditions.

Drinking water must be available. *E. decipiens* drinking.

Every cage should have a water dish or other source of drinking water. Daily mistings with a spray bottle or a tray of water gel crystals are alternative ways to provide water. Water may be poured directly onto a section or corner of the substrate. A water dish can be any shallow container. The inverted jars with a molded water dish underneath (automatic pet waterers) work quite well in large enclosures and seldom need refilling. Cockroaches can drown in open water, so gravel or a piece of sponge foam can be placed inside the water dish to prevent that. Foam is preferred since food or small nymphs can end up rotting between the gravel. Pieces should be cut slightly larger than the dish to provide a tight fit. Sheets of sponge foam are available for crafts and are easily cut with an electric knife or razor blade (scissors, only if it is thin). Foam can get chewed up and may need to be replaced every few years.

The enclosure was kept too dry for adequate hydration and wing expansion, though the *D. paulinoi* survived.

Two things to keep in mind for placement of the water dish:

1. Food should be kept on one side of the tank and the water dish on the opposite side. Otherwise the food can grow molds from high moisture levels. If the food becomes too moldy it will be inedible even for the most voracious roaches.

2. The top of the dish should be the same level as the surrounding substrate.

Watering the substrate in a roach enclosure to provide adequate moisture is like watering plants; it is preferable to dry out a little (never too much) between watering. It may be necessary to water every day during the winter when building humidity levels drop to single digits in temperate areas. Watering the exact same enclosure twice a month can cause rot and kill the inhabitants during the humid spring. The proximity to vents for heat and air in the home can be as important as how the cage is placed back on the shelf. The speed of drying is greatly increased if cage vents line up with the flow of air from structural heating vents. The volume and water content of the food also plays a role. There are too many variables to determine the exact schedule and volume of water needed, so some keepers will invariably have a 'green' or 'brown' thumb. Fortunately, for most species, the learning curve is short and a wide range is acceptable if not optimal.

NUMBER AND TYPES OF ROACHES

Keeping different species together in the same enclosure is sometimes possible. In some cases one species will entirely devour the other, whether it is the other species' oothecae or individuals during molts. Such predation can take place over weeks or many months. *Eublaberus posticus* and to a lesser degree, *P. surinamensis*, are uniquely adapted to killing off other species over time. The majority of other species will live in the same cage in relative harmony as long as there is adequate food and neither builds up to large numbers.

Roaches kept in groups can grow more rapidly or change behavior in other ways (Willis et al. 1958). Overcrowding can be difficult to quantify. Are fifty hissers in a cage too many? What food and water is available? What are the sizes of the enclosure and hides available? If the condition really is overcrowding, it should

Overcrowding in some species is only a problem when an adult molts. The chewed wings on this freshly molted *B. craniifer* resulted more from underfeeding than overcrowding.

lead to measurable problems. These are mostly related to cannibalism. The ends of the wings and antennae are often bitten off, while freshly molted animals can be chewed on or consumed by a hungry mob within minutes. Molting interruptions may cause deformed legs and wings even if cannibalism plays no part. Reduced adult size is a common, less obvious side effect of overcrowding. Some species eat their own eggs, especially all the larger Blaberidae. Ootheca cannibalism results more from limited feeding than overcrowding, but it is less likely

First generation hybrid hisser female, *G. grandidieri* X *G. portentosa*.

in a very large enclosure with many hides and few individuals. Efforts to shield the oothecae from predation such as provisioning the cage with perforated cork bark may be necessary. Remember that overcrowding refers to the negative effects of keeping too many animals in too small of an enclosure, not an absolute number, mass, or weight. In the case of *Macropanesthia*, one male with first instar young is "overcrowded" because he eats them, but after they have molted a few times they can be kept with him safely for years. Larger or more animals in this case do not suffer ill effects from crowding.

When keeping different species together, it is important not to keep closely related species in the same cage to avoid hybridization. Mixing two species, subspecies, or varieties is usually considered a very bad thing, mostly because the resulting stocks are often sold or traded as the wrong thing. Unlabeled hybrids could obliterate the founding bloodlines. Hybrids should be very carefully isolated from founding stocks or destroyed. Species from the same genus are not always able to hybridize. Some documented cases are *Blattella asahinai* and *B. germanica* (Ross 1993), certain *Blaberus* (McMonigle 2008a), and *Gromphadorhina portentosa* with *G. oblongonota* (McMonigle 2008d). Many closely allied specimens like similar *Gyna* spp. or *Therea* spp. have simply not been not been tested for interspecific mating. Some hybridizations may be caused by difficulties of taxonomy, such as inadequate descriptions or the tendency to give different names to different populations of a variable species, since living organisms cannot always be put in a box (literally and figuratively). There is glory and posterity for the taxonomist in splitting, but never in lumping.

The most difficult hybrid problem among captive stocks is only partly caused by the

This is a first-generation hybrid hisser male, "tiger hisser" *G. grandidieri* X *G. portentosa*. The unique pattern is usually lost with the next generation that are brown-black overall.

invalid genus 'Princisia' (McMonigle 2003). Taxonomists have long been aware that 'Princisia' and *Gromphadorhina* male genitalia are identical (Van Herrewege 1973, Beccaloni pers. comm.), and the original description was based on variable characteristics of the male's pronotum. Some *Princisia* stocks are wholly misidentified, but most follow a poorly conceived generic erection. Every hisser labeled 'Princisia' is from the genus *Gromphadorhina,* but this is not the end of the story. Everything labeled 'Princisia' interbreeds with the valid species *G. portentosa* and *G. oblongonota* (and breed with the stocks labeled *G. grandidieri*). In this case it would seem the valid species and the invalid all breed together without any concern for the whims of man's classification schemes. With *Gromphadorhina* the problem is not simply that keepers put the animals together and may not realize or care that they can interbreed. The first instar nymphs are very impressive escape artists and are very good at finding their way into new cages, so hybridizations can happen even when they are guarded against. If a nymph looks out of place, it may be a variant, but it may be from another cage.

Different natural varieties, though they are not identified subspecies, such as the two forms of *Eurytocis opaca*, should not be kept in the same enclosure unless the bloodlines will end there and no hybrids will be passed on to others. In the case of captive isolated variants like yellow *Gyna lurida*, yellow *Blaptica dubia*, or giant hissers, crossing them with the normal form is not hybridization, but they should be kept isolated or the individual varieties will be lost.

PESTS IN THE HABITAT

1) Phorid flies (Family Phoridae) are the most difficult to control large pests associated with captive-reared cockroaches. The maggots feed primarily on fresh rotting insect corpses but will also consume wet dog food, fruit, or fruit fly media. Fruit flies (Family Drosophilidae) are rarely found in roach caging except where large amounts of wet, rotting fruit remain uneaten for long periods. It is easy to tell the two fly families apart because phorids often walk along surfaces and do not have red eyes. Maggots can only be differentiated by their food source, unless the food source is fruit or sugary liquid. Fruit fly puparia have two tubercles at one end (like a little Martian hat) and are usually made very close to the food. Phorid puparia look the same at both ends and are more likely to be away from the food source.

The ability to control phorids is dependent on the size of cultures, species size, ambient humidity, and cleaning schedule. Phorids cannot be easily prevented because they are found worldwide and small specimens walk through regular window screen and tiny structural gaps. Large specimens cannot squeeze through screen, but the same species can develop into very tiny forms depending on how fast the food runs out. Large cockroach cultures and large numbers of cultures are extremely difficult to keep free of phorids because it is difficult to remove every dead cockroach before it cultures new flies. The bulkiest cockroach species are the best food source because hundreds of flies can come from one animal and the moisture level stays high inside the exoskeleton. Low ambient humidity limits phorid numbers and can eliminate them from cultures of small-bodied roaches because the maggots require semi-liquid food. Maggots die if the food dries quickly enough. A daily cleaning schedule will greatly reduce phorids through removal of food sources, but in the case of large cultures of large species, total elimination is more challenging.

Phorid flies can be trapped with commercially available flypaper or baited to a terrestrial version of the minnow trap. Traps can be made using any tightly-sealed plastic container. Small holes are placed in the sides or lid with a small nail or large thumbtack. Holes should be just big enough for flies to enter and a smelly, wet carcass of a cockroach or cricket is used as bait. When the trap is full of flies, the holes are covered with tape and the trap can be tossed out.

Common phorid flies are harmless to healthy cockroaches. The maggots' ability to grow and consume a dead cockroach within days has prompted some fears and confusion regarding cause and effect. However harmless under normal circumstances, if there is massive damage to the body wall of a living cockroach, the maggots can grow and consume the animal alive. I have only seen this once in thirty-five years on an adult *G. portentosa* that had fallen and torn a large hole in its abdomen. The wound scabbed over but the maggots easily ate through the dried blood. Immediate application of liquid bandage or model glue to cover the wound would likely have prevented this.

2) Mites including grain, predatory, and beneficial:

Obligate symbiotic (to the benefit of both the roach and the mite): *Gromphadorholaelaps schaeferi* is the mite commonly found on captive colonies of Madagascan hissing cockroaches (*Gromphadorhina* and *Elliptorhina*). This species is attracted to the smell of *Gromphadorhina* (Schaefer and Peckham 1968) and its close relatives. It is uniquely adapted and refuses other hosts under nearly all circumstances. However, in a unique

Gromphadorholaelaps schaeferi mites on *Gromphadorhina oblongonota* cockroach © Henry Kohler

experiment, this species permanently transferred to a colony of South American *Blaberus* (McMonigle 2015c). These large tan mites are helpful to the cockroaches because they eat harmful grain mites and reduce mold growth. They do not feed on live or dead cockroaches and die if they cannot find a new host. Some of the obligate symbiotic mites found on new, related hissers might be look-alike species. Large mites found on wild *Macropanesthia* and other roaches in nature might also be obligate beneficial symbionts but they have not been studied.

Phoretic: The most widespread, common, and difficult-to-control pests found in cockroach enclosures are grain mites. This is not just a widespread American or European blatticulturist problem, as an apparent phoretic infestation can be seen on photographs of captive *Macropanesthia* in Australia (Rentz 2014 pp. 87-88). The most commonly encountered phoretic mite is *Acarus siro*, but there is also the mold mite *Tyrophagus putrescentiae*, among others. These mites feed on dog food, fruit, vegetables, fresh leaves, and almost anything containing sugar or fat. They do not feed on live cockroaches, but reproduce by the thousands to consume dead ones. Keep in mind, battling grain mites in roach cultures is not a new phenomenon. There have been various methods, including chemical control, tried by blatticulturists dating back at least to the 1940s (Fisk 1951). When food begins to run out, they molt to a hypopus stage which sticks to the exoskeleton of the cockroach. This phoretic stage is designed to cling to flattened areas of living arthropods and if they are scraped off (or if the cockroach molts) they walk right back on. Large infestations can kill cockroaches under humid conditions because the hypopus stage blocks the spiracles and can slowly suffocate the cockroach. Grain mites are tiny and look like fine sand grains that move very slowly without an apparent means of locomotion. Under magnification they appear to move quickly and the legs are easy to see. Grain mites are usually off-white in color.

Ignore the big hisser mite. Look at the dozens of *Acarus siro* phoretic deutonymphs holding onto the indented area of the exoskeleton around the antenna base. © Henry Kohler

Predatory: These tan to brown mites look like smaller versions of the hisser mites, but are still large enough that the average person can easily see their legs without a magnifying lens (common genera are *Stratiolaelaps* and *Androlaelaps*). They feed on grain mites, flies, nematodes and other small invertebrates. They do not feed on even the smallest living cockroach, though they will feed on decomposing

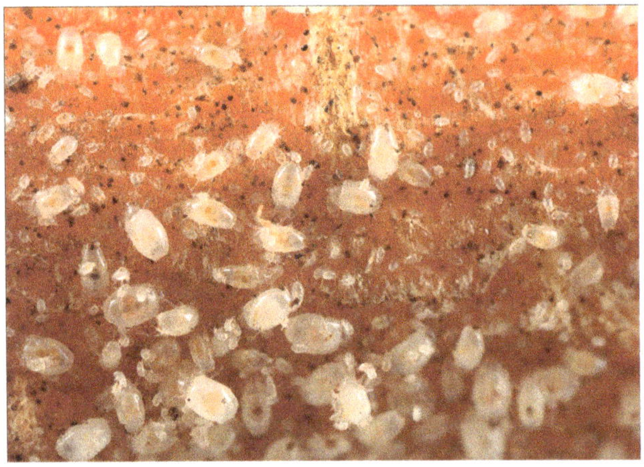

Pest mites on a carrot may be *Tyrophagus putrescentiae*, which are similar to *Acarus* but much more hairy. © Henry Kohler

Small red mites in a cage (with grain mites, white)

cockroaches. Commonly encountered species are not prolific enough to eliminate massive grain mite infestations (only small infestations). However, *Stratiolaelaps scimitus* can be purchased by the thousands from organic pest control suppliers.

Free living: Less common than the above three types of mites are small species that feed on fruits and dog food in the cage (rarely dead roaches). They can be annoying but are harmless to the roaches because they do not have a hypopus phase. These are usually larger than the grain mites and black or red in color.

Parasitic: While other types of mites will almost certainly be encountered by the average keeper, parasitic mites are essentially unheard of in culture. Rarity is likely related to cyclical requirements that would take extra effort to maintain over generations in captivity; wild-caught roaches are not available for any of the popular tropicals. There are parasitic mite species reported from wild cockroaches, like *Blaberpolipus cavernicola* which was found on *Blaberus parabolicus* and *Eublaberus distanti* cockroaches in a cave close to Tingo Maria, Peru (Husband and O'Connor 2003), but reports in captivity are almost nonexistent. However, *Pimeliaphilus cunliffei* has (fifty years ago at least once) decimated laboratory cultures of pest cockroaches used to test pesticides (Field et al. 1966). Other *Pimeliaphilus* mites parasitize darkling beetles, scorpions, and blood-sucking conenose bugs.

3) Drugstore Beetle (*Stegobium paniceum*): These tiny beetles walk right through metal screen. They also fly readily. Drugstore beetles tend to infest old substrate or cages where the dog food sits around for months. They require very little moisture and are sometimes sold and reared as feeder insects, which is probably the most common way they find their way to nearby roach cultures. Since they are tiny it can be difficult to entirely eradicate them, though infestations often die out on their own after a few years. If one piece of dry cat food falls under a piece of furniture, it can support a little colony for a year.

4) Meal moths are pests of stored grain found worldwide. They can infest roach cultures even

G.schaeferi on *G.oblongonota* © Henry Kohler

Meal moths and their wing scales can get trapped on oily climbing barriers and spoil the barrier.

Pyralis farinalis caterpillars can live in the substrate of roach cages.

Pyralis web tubes often connect to dead roaches since the caterpillars are scavengers.

if grains are not used. The least prolific and largest (wingspan up to 30 mm) species seen is *Pyralis farinalis*. Its larvae build large silken tubes near the surface and are covered in dirt and frass. A small (~10 mm wingspan) species, *Plodia interpunctella*, is often found infesting pet foods, but I have never seen it in a roach cage. However, a very similar species is common in roach enclosures. It is not bicolor like *P. interpunctella*, but is silvery gray with darker flecks. Its larval tubes are often deeper and more difficult to find or separate from the substrate. Meal moths are harmless to cockroaches and are useful for feeding small mantids or frogs. Meal moths can be controlled by removing and replacing the substrate once or twice a year.

The adult *Pyralis farinalis* meal moth

5) Fruit Flies (*Drosophila* spp.). The most common species is the laboratory fruit fly, *Drosophila melanogaster*. These flies are difficult to prevent if large amounts of fruit sit around uneaten. This pest is easy to eliminate by switching to dog food for a few weeks or eliminating fruit altogether.

6) Molds and diseases occur in some situations but are not commonly encountered in captive colonies. Helminths, horsehair worms, and flatworms are unlikely to ever be seen by the seasoned keeper. There is a bacterial infection that causes deadly abdominal bloating seen from time to time. Antibiotics designed and dosed by weight for use with small animals may be used, but it can be too late when noticed. Another bacteria results in dead animals turning pink, however it colonizes post-mortem, and is therefore always incurable. It is part of natural decomposition under certain damp conditions, since living animals never turn pink. The pink color is probably caused by the widespread bacterium, *Serratia marcescens*, that is also responsible for slimy pink deposits around sinks and tubs in bathrooms. Of course cause and effect can be confused with poor husbandry and overly damp conditions, but I have seen dead pink roaches in healthy old colonies, still going strong a decade or more afterwards. White or yellow cottony fungus is seen on dead roaches sometimes, but again is normally a product of decomposition.

7) Spiders are usually immaterial when present in a culture, but it is possible for one to consume large numbers of small nymphs. (The number eaten seems directly proportional to the cost and rarity of the cockroach, though the relationship may only be the tendency to notice.) Containers should be closely inspected before being set up to house roaches. Burrowing species are often safe from spiders.

CLEAN-UP CREWS

1) Dermestid beetles reduce problems with phorid flies and the odor of dying cockroaches because they can eat a large volume faster than other decomposers. They can survive and even thrive with little to no water. Unfortunately, these beetles have a slight smell themselves and can be dangerous in numbers since they

Springtails are harmless cage pests.

A dead roach turns pink as damp conditions and food allow *Serratia marcescens* bacteria to flourish.

are carnivorous. Dermestids also feed on dried insect specimens, oothecae, and dry pet food. *Dermestes lardarius* is suggested over the common *Dermestes maculatus* because the adults are relatively slow moving and do not employ flight as an escape response.

2) Isopods: these cut down on molds and smells in damp terraria by consuming leftover food. They are usually harmless to cockroaches and take forever to consume dead cockroaches even en masse. Dry roach enclosures can kill them.
3) *Eleodes subnitens* is a large darkling beetle whose larvae look a lot like the common yellow mealworm (*Tenebrio molitor*). They reduce food available for most substrate pests, but do little for carcass removal and phorid prevention. Adults and larvae are harmless to cockroaches. There is very little effort expended to maintain them as cleaners. Mature larvae must be removed from the cage for successful pupation. A month or two later the new beetle can be returned to the enclosure and will live a few years. One adult pair should produce enough larvae to keep 2″ (5 cm) of substrate in a 10-gallon (40 L) terrarium clean.

ARTIFICIAL DIETS

It is a common myth that cockroaches eat anything. Most cockroaches are omnivores like humans, but they positively do not eat anything. A number of roaches are extremely picky. No roaches are able to digest inorganic materials such as metal or glass. No roach can digest plastic. Only a few species, beyond the termites, are able to survive on a diet of rotten wood alone (and the moisture level, state of decomposition, and type of rotten wood must still fall within a narrow range). The common story of roaches living off of the glue that binds books together holds little merit if one actually attempts to keep their pets alive on this diet. This story likely came about from 1800s food-grade glues that a human could also eat. Roaches do chew on inedible substances when they are hungry and have no choice, but this does not keep them alive. However, most blattodeans thrive on the majority of foods a human or human pet can eat.

Dry kibbled dog or cat food, fish flakes, monkey chow, ferret pellets, etc., are standard fare used to maintain cockroach cultures. These foods offer everything most cockroaches need to grow and can be acquired and stored easily. Rodent pellets for rabbits and guinea pigs are mostly inedible, but rat and mouse chow is acceptable. When I first began keeping hissers in 1980 the standard was monkey chow, but I quickly switched my pets to dry dog food because I did not keep monkeys. Different brands of dried food are variably enjoyed, but any brand can be the sole food for most species. Wet dog food seems to be eaten a little more quickly, but molds a lot more quickly and can invite flies. Moist (soft) kibbled foods are better for damp caging because they resist mold. Some keepers prefer fish food flakes since the smell attracts feeding and the particles are easily chewed, though fish food is relatively expensive and seldom resistant to mold. Cockroach colonies grow and breed at noticeably different rates depending on the brand of dog food, cat food, etc., being fed. Brands with "gravy" and other added flavors are eaten more readily. Protein and calcium can be added to these foods by coating them with powdered milk. (Just shake together in a plastic bag.)

No matter how small the cockroach, whole pieces of dog food should be used. The smallest nymph of the smallest species has no trouble taking a bite off an intact piece of food. Grinding up food is unnecessary and makes monitoring of food consumption difficult. Tiny

Gyna caffrorum feeding on a slice of leftover Halloween pumpkin

Young hissers feeding on a romaine lettuce stalk

Eublaberus posticus feeding on dry dog food

Gyna caffrorum food frenzy over dog food kibbles

particles will sift into the substrate where they are hidden. They rot, mold, or become infested with mites when out of sight. Ground-up small particle foods still work, but they do not add value.

Fruits and vegetables are eaten by most species. Apple, banana, cantaloupe, carrot, gourd, grape, mango, orange, pear, pineapple, potato, pumpkin, romaine lettuce, spinach, strawberry, and watermelon are among the many fruits and vegetables relished by most of the Blattodea. Fruits are usually lopsided in nutrition and if a single food is offered by itself the roaches will get tired of it. Some, like pineapple and potato, can trigger immediate nibbling but tend to be ignored later. With the exception of starving animals and massive colonies, roaches prefer fresh fruit. They will starve rather than eat fruit entirely consumed by mold. However, leftover rinds, cores, and peels are perfectly acceptable. Fruit that is a little overripe is okay, but rotten fruit can be harmful if it is eaten at all. Moldy coconut can kill roaches. Romaine lettuce, rose leaves, and certain other greens are eaten hungrily by *G. grandidieri* and *G. portentosa*, but these foods convert to excessive frass because they are primarily inedible.

Providing fruits and vegetables along with dog food increases the speed of growth and reproduction for most cockroaches. Fruits and vegetables can be used as a singular diet (depending on the species and fruit or vegetable), but they are usually offered along with kibbled dog food because dried foods can sit dry for weeks or months without spoilage. Some of the live-bearing species retain young indefinitely if fruits and vegetables are not offered as part of the diet. *Corydidarum pygmaea*, *Gyna capucina*, and *Opisthoplatia orientalis* are among the species that are unlikely to reproduce without fruit.

There has been limited research into cockroach food preferences and growth. Currently available data is unlikely to provide enough detail to build a methodology (consider the debate over the nutritional requirements of the more popular pet dogs and cats, and those are each only one species). Cockroaches usually tire of foods that lack basic nutrients and their feeding voracity is used to determine the type and volume of food replaced. Nearly all species can be reared quite successfully with dry dog food supplemented with apple slices or other fruit. In the species section advice is given for individual species when diet is at all out of the ordinary. While most species only feed on potato if they are starving (and minimally even then), Connell (2011) performed experiments where *Nauphoeta cinerea* reproduced significantly better on a strict diet of instant potato flakes. Over sixty-four days the young produced by mature females was counted. Alternate groups were provided powdered milk, rolled oats, or the control (dog food, lettuce, and carrots). The adult *N. cinerea* had been reared to maturity on the control diet used at the insectarium. Those fed only compost all died by day forty-five. In other experiments, reproductive female *Gromphadorhina* choose dog food over apple 80% of the time (Carrell and Tanner 2002), but a preference does not suggest fruit is bad for a female. An older experiment questions the most tried and true friend of the blatticulturist, the apple. Hintze-Podufal and Nierling (1986) reared groups of *Blaptica dubia* on parallel diets and the addition of apple was noted to greatly slow the growth of this species. Maturation times for those fed dog food and carrots were 199 days, dog food and cucumber 196 days, dog food and cream cheese 170 days, dog food alone 254, dog food and apples 332. Specimens fed oats and margarine all died before reaching maturity.

Blaberus craniifer feeding on a pumpkin slice

Most roaches, like this adult *Eurycotis lixa,* will feed on honeydew rind.

These *Eublaburus posticus* are swarming over a hawkmoth caterpillar infested with parasitic wasps. Everything except the cocoons is consumed within minutes.

Macropanesthia rhinoceros is among the many species capable of eating wood to some extent, but the wood must be very soft and rotted.

Within reason, the attention paid to feeding behavior and frequency is usually far more important than the food itself.

Various prepared roach foods or roach chows began to show up for sale in the late 1990s. These were not professional formulations, but homemade mixes of ground dog food, chicken meal, rabbit pellets, and other unidentified ingredients. A number of brands have been tested which turned out substandard and generally worse than cheap dog food (according to parameters of acceptance, reproduction and growth) (McMonigle and Willis 2000). On the other hand, prepared beetle jellies made for feeding rhino and stag beetles are a favorite treat for many roaches. Fruit jellies sold for roaches are a treat, not a staple diet.

Fruits, vegetables, dog and cat food, etc., should be offered in a dish whenever possible. Otherwise, little pieces of food and leftovers mix into the substrate and provide a breeding ground for pestiferous grain mites and meal moths. If pests become problematic the cage should be cleaned, feeding reduced, and a feeding dish employed. A feeding dish can be dangerous to small nymphs if they fall in and dry up. The dish may need to be very shallow or include a cardboard "ladder" for escape with species that cannot climb.

Rotten wood is a required food source for a few roaches and may be beneficial for others. Members of the genus *Cryptocercus* are able to digest cellulose due to the presence of symbiotic flagellates in their hindgut. They are found in small family units inside galleries in damp rotting logs and feed on nothing else. Roaches in the subfamily Panesthiinae feed on well-rotted wood and obtain their nourishment from the molds and fungi growing on the wood, but also have foregut bacteria that helps digest wood. Other roaches, including *Panchlora nivea*, *Blaberus giganteus*, and *Macropanesthia*

In crowded cages especially, animals are often seen carrying food away between the front legs to eat in peace.

rhinoceros, feed on decomposing wood and leaves and fare better when these foods are available as part of their diet. When other food sources are not present (or too moldy to eat) the nutrients available in rotten wood may allow these animals to survive to their next meal or even grow. Certainly, no cockroach is able to chew on a fresh or 'green' log, not even termites. However, hardwood sawdust or aspen shavings can be mixed with damp substrate, which grows edible, beneficial fungus in as little as a few weeks.

The blaberid subfamily Geoscapheinae feeds almost exclusively on decomposing leaves in the wild. Polyphagidae also require some dead leaves in their diet and may prefer them to other foods. Rotting leaves should make up all or most of their diet. Leaves should be collected only after they have been outside on the ground for at least a season. Dried green leaves are often ignored. Even brown leaves can lead to massive grain mite infestations if they are freshly fallen. If the leaves have not had some time to decompose they can contain excess sugars and volatile organics.

The speed at which food is devoured is different for different foods, species and colonies.

Feeding on a fruit jelly

Roaches feeding on a mango stone in a gully, Manila Bay, Philippines (CC-SA) Ric_k / Flickr

This male ectobiid is feeding on pollen in a sunlit area outside the rainforest in Panay, Philippines. © Christian Schwarz

This Ectobiidae in Panay seems to be feeding on a spider pellet. Although roaches can be extreme scavengers, the behavior does not always lead to acquisition of useful nutrients. © Christian Schwarz

A few hundred specimens of *Panchlora nivea* eat less dog food in a day than one female *Parcoblatta pennsylvanica*, and still grow just as well. (Conversely *P. nivea* can chew through a lot of rotten wood while *P. pennsylvanica* barely touch it.) When a new colony of *Blaberus* is started it may seem to take forever for the colony to consume a couple of tiny pieces of dog food. When the same colony grows larger, a huge handful can be gone in minutes.

Points to keep in mind: A variety of foods usually increases the rate of growth, survival, and reproduction. Decomposing wood and leaves can be an important portion of the diet. Minimal feeding is normal for small colonies. Mite problems commonly result from overfeeding and not using a food dish. Although huge amounts of food can be consumed, it is amazing how little food each individual actually needs to grow and reproduce.

Eastern box turtle eating a hisser

COCKROACHES AS PREY

Many people who keep roaches as pets started out by keeping them as food for lizards, turtles, frogs, salamanders, tarantulas, mantids, assassin bugs, and other insectivores commonly reared or maintained in captivity. Keeping the Blattodea as pets does not mean that they cannot also be used as prey items. Even those cultures kept solely for observation as pets need to be culled to prevent deformities and small size from taking over the culture stock. Using culls to maintain the health of other animals is an excellent alternative to simply killing runts and deformities. Except for the more difficult-to-breed roaches there must be a way to limit the culture so that it does not outgrow its cage. Alternative methods of limiting culture size, such as reduced temperatures or limited feeding, can harm the colony, whereas feeding off extras is only a problem if attention is not paid to maintaining a minimum required number of individuals.

There are numerous reasons why roaches make better feeder animals than pinkie mice, crickets, mealworms, etc. To begin, the natural diet of many carnivorous pets consists of insects. Both vertebrate and invertebrate insectivores are likely to receive more balanced nutrition if they are given insects instead of pinkie mice (also cockroaches do not scream like mice). Other insect prey, mealworms and crickets, are too small to efficiently feed large predators. Mealworms have limited usefulness as prey because of minimal movement, inability to climb, and a hard shell. Mealworm culturing takes a long time to produce very little, while the substrate is food and often becomes infested with mites. Most people have a very difficult time rearing crickets because of the warmth, attention, and caging needed. Very few people have trouble rearing common feeder roach species. No special substrate or care is needed for egg laying and cages do not require constant cleaning. Purchasing crickets or mealworms can be a permanent cost of housing and

A triceratops beetle (*Phileurus truncatus*) eating a recently deceased *B. craniifer*

Though often much smaller than their prey, the African *Platymeris* and *Psytalla* assassin bugs are among the few invertebrate predators capable of easily dispatching an adult hisser. Much larger tarantulas, centipedes, and scorpions rarely can get past the spines and thick shell.

amount to hundreds of dollars quickly, whereas maintaining a colony of roaches can run a few cents for the dog food and fruit scraps needed to fuel it. Ultimately, the most desirable aspect of using cockroaches as feeders is that, unlike mealworms and crickets, roaches do not eat what they are being fed to. Postultimately, crickets, mealworms, cockroaches, and possibly pinkies can be useful at different life stages or fed in conjunction to offer a varied diet.

The time to set up a culture should be considered. Unless a massive colony is purchased, it usually takes three to six months for a colony to produce animals for feeding. The number needed to start is going to be different based on the number and type of predator being fed, the type of roach chosen, and of course the cost of the starter colony. Rarely can one set up a culture and immediately begin feeding, but once the culture size matches the dietary needs of the predator population it can be used indefinitely. Adult males are the most expendable and can be fed off right away so long as a couple are left behind. One male of most species is plenty for a dozen females. I have removed every male I could find from *N. cinerea* cultures without a drop in production. Small nymphs may also be used as food right away if good numbers are being produced.

Gut loading feeder insects is done in hopes of offering a more nutritious or safer meal for the predator. The food most recently eaten by the feeder cockroach is then ingested by the predator. Gut loading is often unnecessary for invertebrate predators since they tend to discard the stomach lining of the prey (dusting with a multivitamin is suggested). When feeding reptiles, popular advice is to offer feeder animals only fruits and vegetables for a week or so prior to feeding, because the high level of vitamin A in dog and cat food could be harmful. However, common feeder cockroaches tested for vitamin A content lacked an adequate amount to support reptile health (Oonincx and Dierenfeld 2012).

Counting and sifting cockroach prey can be necessary for feeding large numbers of variously sized herps. Sorting is usually done in one of two ways:

A. Large numbers of adults, mostly females, are rotated between cages every moth or so. This way all the nymphs remaining in the cage are of a similar age and size.

B. Sifting trays are made of wire mesh so large groups can be separated accordingly. This is usually only useful with burrowing species like *Blaberus* and *Blaptica* because it can be done quickly over a thin layer of substrate and they cannot hang on.

Many of the easier to breed roaches like *Nauphoeta* and *Blaberus* can be difficult to feed to predators without specific considerations. Burrowing feeders like *Blaptica* and *Blaberus* should be avoided if the predator's terrarium must have substrate. The predator's enclosure will normally have to be carefully crafted or retrofitted to prevent escape of the feeders. A cage designed to hold a bearded dragon or box turtle requires significantly less attention to small holes and gaps than that needed to contain a *Blaberus* or *Gromphadorhina* prey item. *Nauphoeta cinerea* are often used to feed chameleons, but this only works well if they are offered inside of prey cups. Otherwise they would hide in the corners of the cage and never be eaten. A wide, light-colored tray-cup is securely fastened to the caging and the edge is lined with petroleum jelly to prevent escape of the roaches. The lizard picks them out of the tray with its long tongue as it clings to the mesh screening or sits on a branch strategically placed above and in reasonable proximity to the tray.

Yellow orchid mantis nymph eating a freshly molted *Nauphoeta* nymph

This red spot assassin is feeding on a much larger *Blaberus* nymph.

COMMON FEEDER SPECIES

Not every species makes a great meal for every predator. Some cockroaches reproduce too slowly, escape caging predation too readily, are too small, or have excellent defenses against being eaten. There are a number of cockroaches kept as feeders, of which five are the most commonly used. The five most common are *Blaptica dubia*, the discoid roach (*Blaberus discoidalis*), the lobster roach (*Nauphoeta cinerea*), the Turkistan (*Blatta lateralis*), and the Madagascar hissing cockroach (*Gromphadorhina portentosa*). These species can used to feed different types of animals or kept concurrently to offer variety.

Euthyrhyncus floridanus feeding on *B. dubia* nymph

1. *Blaptica dubia* is probably the most common feeder today. I have even seen this species for sale in the feeder section at the big box pet stores. They are kept in huge numbers around

Euthyrhyncus floridanus feeding on *S. lampyridiformis* nymph

the world as food for reptiles, amphibians, and small mammals such as pygmy hedgehogs and monkeys. *Blaptica* are used as a major component of the diet for conditioning a variety of breeding reptiles. First instar nymphs are small enough to feed small frogs and lizards, while adults are big enough to provide a hearty meal for a monitor or turtle. They do not climb. Overall this species is similar to *Blaberus* and *Eublaberus* as feeders, but the adult females do not have large wings and the exoskeleton is a little less rigid. The main difference is that *Blaptica* can breed faster at higher temperatures, but is more demanding because it needs 77-86° F (25-30° C) to reproduce well.

2. *Blaberus discoidalis* was once the most commonly available feeder. It does not require high temperatures and can reproduce reasonably well down around room temperature 72° F (22° C). Nymphs and adults, like most of the

feeders, possess a mild defensive odor which does not deter commonly kept predators. The different sizes available for feeding range from about an eight of an inch (8 mm) at birth to nearly two inches (45 mm) at adulthood. This roach is unable to escape from smooth sided cages and so is usually easy to contain during feeding. Discoid cockroaches are still easy to acquire and among the easiest to rear.

3. The lobster roach (*Nauphoeta cinerea*) is another common feeder, but it is a glass climber. Nymphs can be used as food for smaller or "picky" eaters. The exoskeletons of *N. cinerea* are thinner and softer than those of crickets or mealworms. The wings of the adult are short and thin, which makes them easier to ingest. Lobster roaches are fast breeders, easily cared for, do well in small rearing cages, and cultures rarely crash. First instar nymphs are barely a few millimeters in length and are excellent prey for hatchling frogs, lizards, and tarantulas. They make good prey for small mantids and whipspiders as soon they have outgrown fruit flies (*Drosophila*). The lobster roach does not make great food for really large reptiles because it would take six to ten adults to make up for the mass of one adult of the larger feeder roaches.

4. *Blatta lateralis,* Turkistan roaches or red runners, are spectacular prey items for small predators. They are smaller and frailer animals than other common feeders, and tiny nymphs are very difficult to handle without them being smashed. Feeding is more successful if they can be shaken from a container than picked out by hand. They do not hide as easily from predators as other common feeders, because they do not burrow and do not squeeze flat into narrow cracks and corners. Of course they do hide in dark crevices if provided places to hide.

5. The widely popular *Gromphadorhina portentosa* has been used to feed many things, though almost never in quantity. The adult hissing roach has an extremely thick and tough exoskeleton. Despite this fact, or rather because of the hard shell, many monitors, turtles, and other large insect-eaters really enjoy munching away at these crunchy critters. Box turtles tend to be very slow and secretive animals, but they literally come out of their shells after they become accustomed to eating hissing roaches. The turtles run each other over and fight over the delicious, crunchy roaches. Most large invertebrate predators like the largest tarantulas and centipedes are intimidated by the size and thick shell. When it comes to invertebrate predators, only the relatively smaller *Platymeris* and *Psytalla* African assassin bugs readily dispatch adult hissers. Young *G. portentosa* nymphs are small and soft-shelled and can be used to feed small animals, but are not produced at the speeds of the preceding species and run out quickly. Early instar hissers easily escape common reptile enclosures.

The above list is not meant to be exhaustive, as there are dozens of species used as feeders. *Schultesia lampyridiformis* is a valuable feeder for whipspiders and scorpions. The green adults of *Panchlora nivea* have been used to elicit feeding responses from animals that are sickly or have given up on feed. *Eublaberus* sp. 'Ivory' were an important feeder in the herp hobby associated with the first time successful captive breeding of some lizards. Even the beautiful *Therea* and massive *Archimandrita* have been used as a source of feeders from time to time.

Roaches almost always have some chemical defense, but most defenses deter few predators. Some cockroaches emit surprisingly

Vinegaroons eat any type of cockroach; this *Typopeltis* is feeding on a pallid roach.

Sparassidae eating an ectobiid, Panay, Philippines © Christian Schwarz

Hemidactylus platyurus eating a roach in Pandan, Panay, Philippines © Christian Schwarz

Gecko gecko vs. roach in a house in Panay, Philippines (Some tropical roaches are listed as pests because they are commonly found in homes or huts. The designation seems unfair when lizards (and sometimes monkeys) are running around the same houses.) © Christian Schwarz

White-faced capuchin eating a giant cockroach, Costa Rica (CC) Jaan Altosaar

Scorpion (*Lychas* sp.) eating cockroach, Malaysia (CC-SA) Bernard Dupont

strong offensive smells (specifically *Eurycotis* spp. like the Florida skunk cockroach) when attacked, and would not seem to make good feeders for anything. However, arthropods such as arachnids, mantids, assassin bugs, and giant water bugs do not seem to mind at all. Egg-laying roaches are normally able to reproduce more quickly than the fastest live-bearing types. Consequently, they can be an excellent food source for the voracious appetites of carnivorous arthropods. While the majority of cockroaches make excellent feeders to supply the needs of insectivores, there are invariably some black sheep. *Deropeltis paulinoi* and *Diploptera punctata* produce noxious quinone defensive fluids that stain the keeper's hands brown during handling. Very few animals attempt to eat them. Even normally non-selective invertebrate predators avoid them or learn to avoid them unless starving.

The Blattodea are an inexpensive and easily kept feeder item for many reptiles, amphibians, invertebrates, insectivorous mammals, and birds. Other commonly kept animals that eat cockroaches, such as rats, mice, and cats are unlikely to ever be given these feeders as they are unnecessary. Some fish keepers rear cockroaches for food, but the most common story is that of archer fish (*Toxotes* spp.) making a meal of circumstantial structural pests. Many commonly kept predators seem to be happier and more energetic after feeding on some cockroaches. Giant roaches provide a huge meal packed with vitamins and protein. While still on the discussion of food, the human digestive system resembles other naturally insectivorous primates, so a good portion of cockroaches is likely much healthier than eight ounces of red meat. So, please save some food for your pets.

Fried insect vendor in Bangkok (CC-BY) Vasenka Photography

Emerald cockroach wasp (*Ampulex compressa*) egg and larva (top to bottom) on *P. americana* © Tibor Ludányi

AS HUMAN FOOD

Chinese medicine is probably the number one way people eat roaches, in terms of thousands of pounds (more than incidentals in agriculture). The Chinese medicinal cockroach (*Eupolyphaga sinensis*) has been reared for this purpose for hundreds, if not thousands, of years. However, in the last decade the demand has picked up. Farmers have switched to *Periplaneta americana* to get the fastest turnaround to meet the demand. This change begs the question of what value, if any, this cure offers, since *Periplaneta* and *Eupolyphaga* are very different in the chemicals they produce.

Cockroach-eating contests at places including Six Flags amusement parks have been somewhat popular in recent decades with the food of choice being *Gromphadorhina*. Game shows like *Fear Factor* and *The Last Passenger* also offer prizes to those who can eat the most live cockroaches in a sitting. Most local contests died out for a few years after a man in 2012 died following a contest at a pet shop. He did not have an allergic reaction as first reported, but choked.

SPECIALIZED COCKROACH PREDATORS

The emerald cockroach wasp (*Ampulex compressa*) is kept by some insectariums and hobbyists as a pet (Ludányi 2012a). Roaches must be reared continuously to fuel its reproduction and survival, since the wasp only lives two to three months. An adult female catches an adult American cockroach, stings it in the brain, and lays an egg on it. Over forty-five days the larva consumes the cockroach innards, pupates within its empty exoskeleton, and emerges as a large, metallic-green-and-blue wasp. There are more than a hundred *Ampulex* species which specialize on various cockroach species, but most are not so impressive. *Ampulex canaliculata* from Canada and the U.S. is a black-colored species, about a third of an inch (~8 mm) long, which specializes on native *Parcoblatta* cockroaches. Various tiny species of Evanid wasps whose larvae feed on cockroach oothecae have been reared in labs, not as pets, but in hopes of controlling pest roaches.

Emerald cockroach wasp (*Ampulex compressa*) © Tibor Ludányi

The oriental ensign wasp, *Evania appendigaster*, is a parasitic wasp that lays eggs inside roach oothecae, which are consumed when the larvae hatch. (CC-BY) A N Suresh Kumar

Shiny Perisphaeriinae from Singapore © Melvyn Yeo

A pill cockroach, *Perisphaerus* sp., from Singapore © Melvyn Yeo

A spotted roach from Singapore © Melvyn Yeo

Nymph showing defensive secretion in Bolivian Amazon © Arthur Anker

SPECIES ACCOUNTS 物種統計

Adult length represents the distance from the front of the pronotum to the end of the wings or abdomen, whichever is greater. Growth periods and adult lifespan are primarily reported for room temperature, around 70-75° F (21-24° C). Times can be affected by higher and lower temperatures, food, moisture, and ventilation. Culture dates are oldest verifiable through literature or personal documentation. Many stocks are likely years or decades older. The American Cockroach Society number is assigned by species or subspecies and does not change even if the scientific name is revised.

Aeluropoda insignis Butler 1882
Flat Horn Hissing Cockroach, ACS #91
Adult length: 35-66 mm
Glass climbing: Yes
Reproduction: Live-bearing
This is one of the newest and most unique hissers to enter the hobby. Adult males have notable "horns" but they are flattened and knob-like. Their unique feature is the large indentation in the pronotum which projects forward on each side at the front. *Aeluro* means 'cat' while *poda* is 'foot'; *insignis* means 'designated' or 'marked.' What part of the body resembles a cat foot is uncertain, but it does not refer to the tarsi because its feet are no different from other hissers. However, the set of

Aeluropoda insignis male

Aeluropoda insignis female

Aeluropoda: on a large male from underneath, the pronotum sculpturing almost looks like a pair cat paws holding onto the edge of the pronotum, on each side of the head.

bumps at the end of the projection on each side of the male's pronotum look like a pair of tiny cat paws holding the edge. Females have no physical features that easily discern them from *Gromphadorhina*. They are a handsome, dark-and-light-patterned maroon like the males.

Adult size is rather variable in captivity, and also apparently in the wild. When Butler described this species in 1882, he also named *Aeluropoda gigantea*. However, he based that description on size alone and later examination of 'gigantea' specimens showed they were simply large *A. insignis*. There is only one species in the genus *Aeluropoda*. Adults are normally smaller than common hissers at two inches or less (<50 mm). However, large colonies produce a few supersized male specimens at close to three inches (66 mm) (McMonigle 2011a). Unlike other hissers, supersized females are rare if they exist at all. Gigantic males are spectacular, although they do not possess an allometric or major horn form like the related *G. portentosa*. With allometry you can get an idea of the size of a male from the shape of the horns, but with *A. insignis* the horns are relatively flattened at any size. However, there is allometric progression related to the massive notch at the front of the pronotum and the points on the 'cat paws.'

While the various *Gromphadorhina* (including 'Princisia') found in the hobby have proven to readily interbreed, it is likely *Aeluropoda* cannot because of differences in the male genitalia. They are housed and fed like other hissers. Specimens were very rare and expensive 2006-2009, but have come down considerably as they breed nearly as readily as the common *Gromphadorhina portentosa*.

Anallacta methanoides

Anallacta methanoides Shelford, 1908
Mauritian Zebra-Faced Cockroach, ACS #108
Adult length: 18-21 mm
Glass climbing: Yes
Reproduction: Egg-laying
Found: Mauritius Island, east of Madagascar

Archiblatta hoeveni Vollenhoven, 1862
Hoeven's Cockroach, ACS #92
Adult length: 45-55 mm
Glass climbing: Yes

Aeluropoda insignis male

Anallacta methanoides

Anallacta methanoides

Arenivaga sp. from Arizona

Archiblatta hoeveni female
© Tanaka and Komatsu

Archiblatta hoeveni male
© Tanaka and Komatsu

Archiblatta nymphs
© Tanaka and Komatsu

Reproduction: Egg-laying
This large and unusual species is a native of Malaysia and Indonesia. The large size, strange body shape, and gigantic, banded antennae have made it a sought-after species. The males are much thinner than the females and fully winged, but can still be shorter overall. Both genders have thick, colorful antennae longer than the body.

Survival of young to adulthood is generally low and the species has proven difficult to keep. *Archiblatta hoeveni* has been kept in Europe and the United States, though cultures may only currently be surviving in Japan (if anywhere). They were being reared by Japanese hobbyists possibly only as far back as 2012 (pers. comm. Hideki). Oothecae contain a dozen to twenty eggs and female produce around ten in a lifetime. It takes about three months for oothecae to hatch at an ambient temperature of 77-86° F (25-30° C). Nymphs that survive can grow to maturity in another five months (Komatsu 2009).

Archimandrita tessellata (Saussure, 1893)
Peppered Roach, ACS #3
Adult length: 55-75 mm
Glass climbing: No
Reproduction: Live-bearing
The peppered cockroach is the heaviest blaberid after *Macropanesthia*. Females have a rounded, humped appearance to the massive body, while males are flattened with a shape similar to *Blaberus*. From above and in many photographs, the two genders may look similar, but the difference in shape makes it easy to differentiate adult genders. (Total length is useless to determine gender though subadult females are easy to pick from subadult males by mass.) The peppered spots on the wings can be very light or very heavy and are not gender dependent.

Archimandrita tessellata pair, female on right

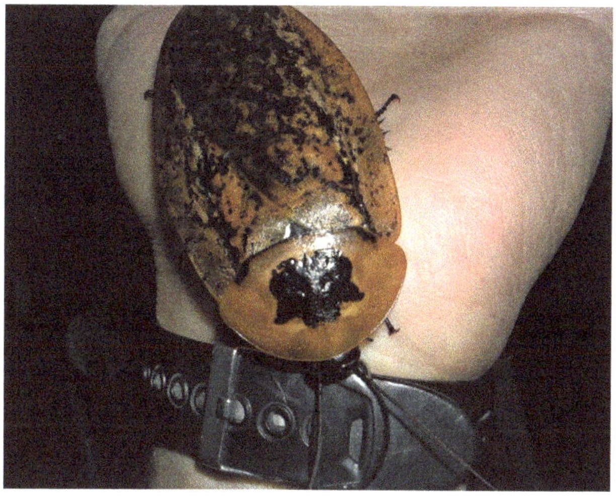

Archimandrita tessellata

Gestation of two to three dozen eggs inside the female takes six months. Nymphs take nine months to become adults, staying buried in the substrate for much of that time. They can eat a large quantity of dog food and fruits for their mass. Peppered roaches live one to two-and-a-half years after reaching adulthood, with males dying first. Old females (>20 months) usually will not reproduce and their tarsi tend to fall off.

This species is found mostly in Central America from Guatemala to northwestern Columbia. The BCG newsletter 2.2 (1987) mentions this species had already been in culture for a while in the United States and that U.S. members "seem to have been taking cockroach culture more seriously and for a much longer time" than European keepers. Current stock is decades old and European bloodlines came from USA stock in the late 1980s, so there may be a single origin for this species. Sometimes this same stock has been traded as *Archimandrita marmorata* (Stoll, 1813), but that species has probably never been in culture.

Arenivaga floridensis Caudell, 1918
Florida Sand Cockroach, ACS #2
Adult length: 12-17 mm
Glass climbing: No
Reproduction: Egg-laying
Adult females are small, wingless, circular creatures with a shiny notum. Males are fully winged. Adults are usually dark brown to black, but females can be mottled tan with darker markings like nymphs. They live in sandy areas where old leaves accumulate and nymphs can be numerous. *Arenivaga floridensis* is the only species in the genus known from Florida. They feed on decaying leaves and animal or plant matter that accumulates in dry scrub areas. Females live two to three years, while males live only two months. Although Florida is humid and damp, *A. floridensis* come from dry areas and will not live long if the substrate is constantly kept soggy.

An additional forty-seven similar species of *Arenivaga* are found in dune and sandy areas across much of the southwestern U.S. (Arizona, California, Colorado, New Mexico, Oklahoma, and Texas) and throughout Mexico (Hopkins 2014). They are differentiated mostly by small variations in the genitalia since coloration, size, and most external features are variable within each species. Most are as small as *A. floridensis*, though *Arenivaga grata* Hebard

Arenivaga bolliana from Charlotte, Texas © Alan Jeon

Large *Arenivaga floridensis* nymphs from Lake Placid, Highlands County, Florida © Alan Jeon

Arenivaga floridensis, subadult male © Alan Jeon

1920 males may reach 27 mm. Many species are known only from the males, since they fly and come to lights. Even when the female of a species is known, the specimens are usually indistinguishable. The most common place to find these species after lights is inside burrows of desert wood rats, kangaroo rats, and prairie dogs. Nymphs can be common under rocks, at least in the town of Tortilla Flat in the Superstition Mountains of Arizona (pers. comm. Clausen).

I kept a colony from 1998 to 2007, of what were probably *Arenivaga apacha* from Arizona. A few dozen nymphs had been collected from a depression where leaf litter, mostly mesquite, built up and mixed with the sand. Collected along with them were a few dozen adventive *Porcellio* sp. (a surprisingly dry-tolerant isopod). These western *Arenivaga* specimens were reasonably easy to keep, but very slow growing for something so tiny. Very little interest was shown to foods other than decomposing leaves. Even when the culture had built up to hundreds in various stages of development, it was unlikely to eat a ¼″ chunk of dog food in six months. Small fruit slices offered only seemed to disappear when they rotted away. The food provided consisted of decomposing mixed hardwood leaves from a northeastern forest. A handful of leaves lasted the colony six months to a year.

Blaberus atropos (Stoll, 1813)
Atropos Cockroach, ACS #5
Adult length: 38-48 mm
Glass climbing: No
Reproduction: Live-bearing

Blaberus atropos is the most variably colored of the *Blaberus* species listed here, at least within a single culture. It seems a little ironic because in Greek mythology, Atropos was the oldest of the fates, known as the inflexible or unchanging one. It is a wonder if this name was supposed to refer to any aspect of this animal's biology. The coloration of most individuals is definitive of this species with a very large, shapeless central splotch and large dark markings on the shoulders of the tegmina. However, about twenty percent of adults resemble *B. discoidalis*. Beyond male genitalia dissection, this species can be identified by the variable adult coloration, pattern of dents on the pronotum, and because nymphs appear similar to *B. discoidalis*, but have two white spots on the head instead of four (normally).

Blaberus atropos, Florida stock
© Tristan Shanahan

Specimens do well on fruit and dog food and usually take six months to reach adulthood. Adults live five to ten months with the males dying first.

This species is from northern South America with the holotype from Guyana. Natick laboratory stock 1969 probably originated in Trinidad. Early 1990s stock was in captivity for an unknown number of years and its locale unknown, and may have derived from the Natick stock. Current stock was collected a few years back in Florida, USA (Kandilian 2011).

Blaberus boliviensis Princis, 1946
Bolivian Cockroach, ACS #6
Adult length: 40-55 mm
Glass climbing: No
Reproduction: Live-bearing
This sturdy creature, as the species name suggests, hails from Bolivia in South America. It is among the most recent members of the genus to show up in the hobby, as recently as 2004. The identifying external feature is a pair of dark brown spots in the middle of the black spot centered on the pronotum. At a glance the adults greatly resemble the discoid cockroach *Blaberus discoidalis*, but under bright light or a photographic flash the double dots stand out. The feature is easy to see but also easy to miss.

Blaberus boliviensis

The Bolivian is a member of what may be the most popular cockroach genus and as such is a live-bearing, nonclimbing species with cryptic, burrowing nymphs. The nymphs are marked in brown and tan and are probably indistinguishable from *B. discoidalis* upon even the closest of examinations. This is among the hardiest of *Blaberus* and adults require little attention and no special stimulus to produce young. Also, adults do not experience mysterious die-off associated with some of the larger *Blaberus*. It is a unique species and makes a great feeder.

Blaberus colosseus adults, Peru

Blaberus colosseus (Illiger, 1801)
Colossal Cockroach, ACS #7
Adult length: 50-78 mm
Glass climbing: No
Reproduction: Live-bearing
This is the second largest *Blaberus* species after *B. giganteus*. Common stock can be easy to identify due to its large size and minimal markings on the tegmina. Nymphs are very dark, almost black in color, which makes them stand out from other *Blaberus*. Adults live six to twelve months. Males are notably shorter and smaller than females and can usually be identified from above.

This species has a wide range from southern Mexico through Central America into Peru and Venezuela. There are two current stocks. The large, easily identified stock lacks dark markings on the tegmina and originated from Peru late 1990s (McMonigle and Willis 2000).

Blaberus boliviensis

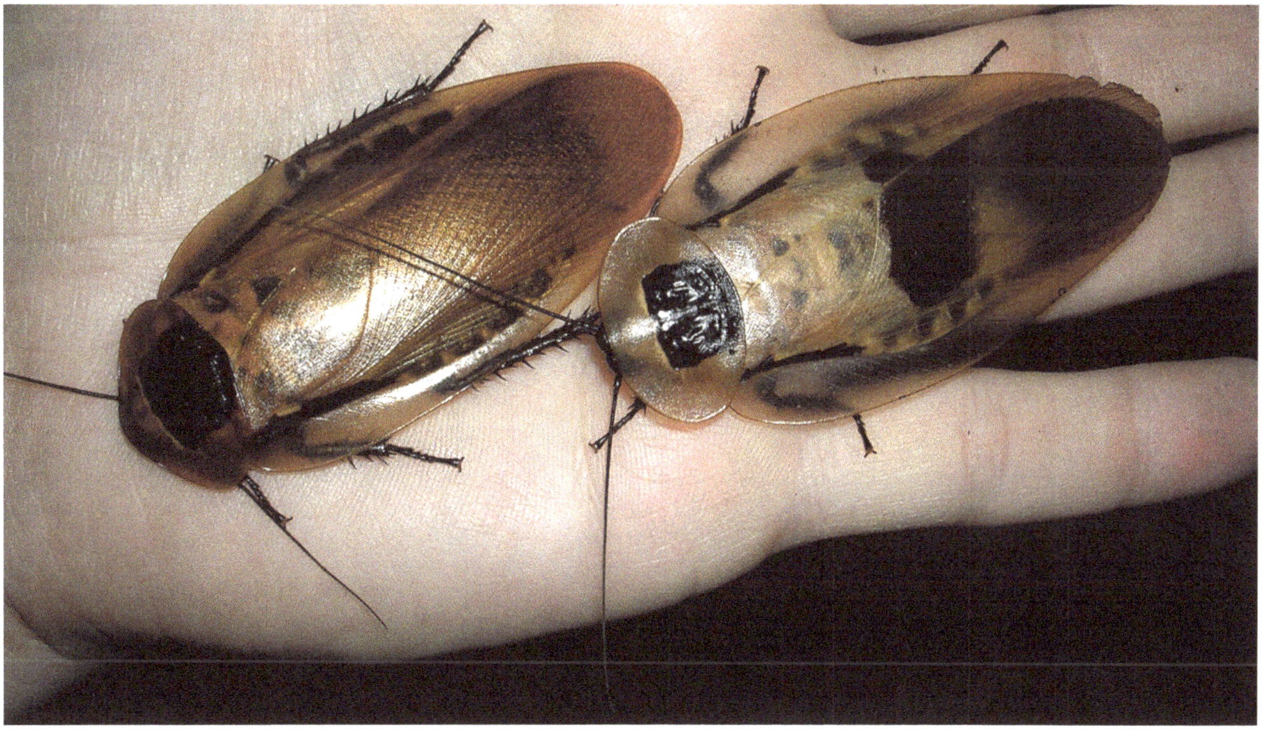

Blaberus size comparison: female *B. giganteus* and female *B. colosseus*

Blaberus colosseus nymph coloration, Peru

A smaller stock from Ecuador (circa 2005) is marked like a small *B. giganteus* and is only differentiated by tegmina setae, nymph coloration, and male genitalia.

Blaberus craniifer Burmeister, 1838
Death's Head Cockroach, ACS #8
Adult length: 45-60 mm
Glass climbing: No
Reproduction: Live-bearing

Blaberus craniifer is a spectacularly beautiful animal and the most recognizable member of the genus *Blaberus*. The death's head cockroach sports a pattern on the pronotum that resembles a skeleton face or "face of death." The name *craniifer* means 'cranium bearing.' The adult female's pronotum has an overall dull, granular surface (unlike other commonly kept *Blaberus*), is slightly upturned, and is proportionately larger compared to the males. Nymphs are variable in color, mostly dark brown, and less distinct.

This is one of those peculiar insects whose scientific name and common name have both encountered a bit of confusion. *Blaberus giganteus* were labeled *Blaberus craniifer* in at least one popular book (Stanek 1969). *Blaberus discoidalis*, a much smaller species in the same genus, was commonly traded under both scientific and common names for this species in the late 1980s and 1990s even though that species has no "death's head" markings. Someone likely knew the genus, ran across the name, and thought it sounded appealing. It was much too small and looked nothing like the line drawing of the real animal from the Florida Keys found in Helfer's guide (1963). The dark-winged form of this species had been called *Blaberus trapezoideus* (pers. comm. Roth 1999).

The real death's head was something I had searched for over the years, but it did not exist in the hobby. I followed lead after lead, always finding discoids. I did not come across specimens until 1998—amazingly it was from a major university just twenty miles from my house. I was doing unrelated work there and heard about the 'cockroach room' and had a chance to check it out. I cannot describe the feeling of excitement when I saw two 55-gallon (208 L) trash cans full of these beautiful animals. Here they were in person with jet-black wings and a scary, little "jack-o-lantern" face on the pronotum.

I am relatively sure that every real death's head in the hobby originally came from this stock as the black death's heads did not exist in the European hobby prior (McMonigle 2008b). The origin of the pictured stock is listed as Florida and may date back to the 1970s (Bell 1981) or even earlier (Wedberg et al. 1949). Another stock, from the University of California Riverside, showed up a few years back and is slightly different but may be of the same origin. This is a widespread species found in Mexico, Central America, and Key West, Florida, U.S.

Around 1999 I gave some to Richard Willis, who happened to also get something in culture

Blaberus craniifer adult and nymph

Blaberus discoidalis

from Europe labeled *B. craniifer*, and he threw them both in the same cage. There are a number of different *Blaberus* species and except for the real death's head the coloration is not especially distinct for any species, so male genitalia dissection can be the only certain way to determine many species. Difficult identification and lack of documentation are a few of the reasons why there is no way to know what was crossed with these to produce the hybrids. Despite some differences in male genitalia, *Blaberus* species can hybridize but so far just with closely allied species (such as *craniifer-fusca-giganteus* or *atropos-discoidalis-boliviensis*). It might have simply been a different geographic race of *B. craniifer* that were also being traded in Europe as *B. fusca* (oddly a listed synonym of the much smaller *B. atropos*, a *B. discoidalis* look-alike species).

All the hybrids (whether interspecific or subspecific) in culture came from R. Willis originally, who sold them to many keepers. The hybrids seem to have retained the higher breeding rate of the '*B. fusca*' but would throw dark- or black-wing forms now and then. The death's head part of the cross is not as fast breeding, so the coloration may eventually die out of the hybrid cultures.

The death's head's geographic color race was originally described at *Blaberus trapezoides*. It was later synonymized by Dr. L.M. Roth with *B. craniifer*. As mentioned, this color pattern is unique to this species, but the species is not specific to the color form (no other *Blaberus* can look like a death's head, but geographic races of *B. craniifer* may look similar to other *Blaberus*).

Like all members of the genus *Blaberus*, it cannot cling to smooth glass or plastic and does not require a culture container lid. It is live bearing and adults rarely live more than a year. Males often die two to three months after reaching adulthood, while females may survive eight months or more. The death's head breeds fairly well and is not prone to mysterious deaths like *B. giganteus*. However, like *B. giganteus*, it should be provided upright molting surfaces for correct wing formation. While rapid reproduction is possible, it does not breed as quickly under as wide a range of conditions as common feeders like *B. discoidalis* and *B. fusca*. Although it can reproduce as low as 68° F (20° C), 82° F (28° C) is the optimal breeding temperature. Dog food supplemented with occasional fresh apple slices is a good captive diet.

Despite nomenclatural issues, there is visibly only one death's head cockroach and its spectacular coloration and large size have made it one of the most desirable pet species.

Blaberus discoidalis Serville, 1838
Discoid Cockroach, ACS #9
Adult length: 35-45 mm
Glass climbing: No
Reproduction: Live-bearing

Blaberus discoidalis is among the smaller members of its genus and may be the least unique. Nymphs are light tan with brown-checkered markings. Gestation period is about eight weeks. Adults live six to ten months, though some females live longer. Nymphs grow to adulthood in as little as four months. I am told cultures can survive up to a few months in a closet without any attention.

The discoid cockroach was the most commonly reared feeder species in the late 1980s and 1990s, and for at least fifteen years this was the only *Blaberus* that could be acquired from vendors. Current culture stock may have come from Jamaica circa 1950 (John and Lewis 1958) and has long been available from scientific supply houses (Carolina Arthropods Manual 1982). This species may be found across the West

Blaberus discoidalis

Blaberus fusca adults and subadult nymph

Blaberus fusca, various sized nymphs and adult, with a stray hisser

Indies, Central and South America, Mexico, and the Florida Keys (Helfer 1963). However, I am not aware of any documentation of this species collected from the Keys (just *B. atropos* and *B. craniifer*).

Blaberus fusca Brunner, 1865
Dusky Cave Roach, ACS #10
Adult length: 50-68 mm
Glass climbing: No
Reproduction: Live-bearing
Blaberus fusca has the narrowest adult shape of the common *Blaberus* species. Like *B. giganteus,* the black pronotum splotch often has two or three white spots in the middle, but rarely, if ever, the cranium markings. Also, like *B. giganteus* the pronotum of the adult female is similar in relative dimensions to the male and mostly smooth and glossy (unlike *B. craniifer*). The markings on the tegmina are similar to *B. giganteus*, but with a noticeably larger and more solid dark band across the middle. Nymphs are similar in color to those of *B. craniifer* when small, but gain some reddish maroon pigments in the last two instars, leading to some confusion with *B. giganteus.* Like *B. giganteus* and *B. craniifer*, it is possible to see golden setae on parts of the tegmina when viewed from the side. Some taxonomists consider *B. fusca* a form of *B. craniifer* due to similarities in the male genitalia. However, the different coloration, shape, and pronotum structure of this population set it apart. I sent some to L. M. Roth in 2001 and explained the differences. He told me the setae on the tegmina suggested *B. craniifer*, but he did not rule out a separate species identification and mused about what other *Blaberus* it could mate with, if any. Since he was the taxonomist who synonymized *B. trapezoideus* (the black wing) and *B. craniifer*, maybe he was just being diplomatic.

About six months are spent as nymphs, while adults live an additional six to twelve months. Females breed very easily with little care, but should be kept at least 73° F (23° C) or the mothers stop giving birth. Current stock is extremely hardy, productive, and requires very little variety of food.

Saupe (1928) documented herding and maternal care of this species under laboratory conditions that were later attributed to *B. craniifer*. The change in attribution was based entirely on his drawings of the pronotum (not genitalia or tegmina features). Late 1980s stock was of uncertain origin, but it was probably not related to Saupe's creatures. This stock was also seen for sale from a French company called Cetonia in the mid-1990s. It may have also been the bloodline listed on the 1997 BCG culture list.

Blaberus giganteus (Linnaeus, 1758)
Giant Cave Cockroach, ACS #11
Adult length: 73-90 mm
Glass climbing: No
Reproduction: Live-bearing
The giant cave cockroach may be the most spectacular of the pet roaches and is the longest winged species kept in captivity. The impressive half-a-foot (153 mm) wingspan placed it squarely at the top of the list of dream species for early hobbyists. Adult males are notably smaller than females and range from 74-78 mm, while adult females measure in at a whopping 82-90 mm (though despite rumors, the maximum verifiable record for any roach specimen is just 97 mm). Females above 86 mm are exceptional but show up now and then in large cultures. Still, even a runty 82 mm female has a wingspan of 151 mm.

Some writings from the early 1960s mention keeping *Blaberus* as pets but do not specify the species. However, *B. giganteus* is seen

Subadult comparison: *Blaberus fusca* (left) and *B. giganteus* (right)

Blaberus giganteus

Blaberus parabolicus, adult female

mislabeled in some earlier entomological texts as *B. craniifer* (Stanek 1969). In the United States this species has been kept as a pet since at least 1972 (McMonigle 2008a) and earlier as laboratory animals (Roth and Willis 1960, Banks 1969). Livestock was extremely difficult to acquire prior to the mid-1990s. Although easily acquired today and inexpensive, it is still arguably the largest and most spectacular pet cockroach.

The giant cave cockroach is a very beautiful species, which grows and breeds much slower than other *Blaberus*. Reaching adulthood takes over a year for the majority of culture stocks. Adult males can fly reasonably well, although adult females grow too heavy soon after becoming adults. Males are among the more aggressive *Blaberus*. They fight and can tear each other's legs off sparring over a female. This behavior subsides with larger numbers of males and not all males are so aggressive. Adult males live from one to six months after the final molt and females seldom live over a year as adults. Up to a quarter of the adults fall over dead within a day of the molt no matter how good the conditions are. If there is limited food or water, every newly emerged adult can die.

Nymphs and adults do well on a diet of 60-80% dog food supplemented with fruit. Unlike other *Blaberus,* they tend to have no births if fruit is entirely lacking from the diet. 72-80° F (22-27° C) is a good range to maintain a colony. Drinking water should always be available. With attentive care they can reproduce in numbers, but neglect can easily wipe out a colony. This species is rarely used (or at least not for long) as feeders because the level of required care is higher and the reproduction level much lower than other *Blaberus*.

The *Blaberus* genus is one of the most popular and established genera in the roach hobby and some mix-ups have occurred. In the 1990s it was common to purchase '*B. giganteus*' and find out later you were sold the relatively tiny *Blaberus discoidalis*. Today mislabeled specimens are less common but now and then *Blaberus fusca* are sold as *B. giganteus*. *Blaberus fusca* are much smaller but are similar in appearance, even somewhat as nymphs. Adult *Blaberus colosseus* 'Ecuador' has similar adult markings to *B. giganteus* but with very smooth tegmina.

Some mislabeling is certainly accidental but it seems smaller, less-valuable creatures tended to be mislabeled as more expensive ones, never the other way around. Some hybrid *B. fusca?* x *B. craniifer* stocks are mislabeled as '*B. fusca* x *B. giganteus*' or an extra specific name is used. The only *Blaberus* species that is nearly as big (might even confuse a ruler at the low end of the range) is *B. colosseus* 'Peru' but the adults look different and only *B. giganteus* has the trademark orange-maroon colored nymphs.

Giant cave cockroaches are found in Central and South America and the West Indies. Colonies are known from bat caves, but they are more commonly found in much smaller caves, in hollow stumps and under logs in forested areas. It is hard to say if the current stock dates back to the 1960s, but certainly at least to the 1970s, while its natural origin is unknown.

A common university strain (I first saw these from Purdue 1996) is notably smaller (seldom reaching 80 mm) but breeds very easily and has overpowered the old strain in culture. Even this smaller, hardier strain is plagued by premature die-off of adults. A visibly narrower strain collected from Trinidad has been at the Cincinnati Zoo since some time prior to 1992.

Blaberus parabolicus Walker, 1868
Peruvian Black Head Roach, ACS #12
Adult length: 50-60 mm
Glass climbing: No
Reproduction: Live-bearing
The Peruvian black head roach appears similar to *B. discoidalis*, but adults are measurably larger and darker. The abdomen is mostly black and the dark mark on the pronotum is larger in relation to the size of the pronotum than for other *Blaberus* except for some of the variable *B. atropos*. *Blaberus parabolicus* is among the faster reproducing and is probably the longest lived member of the genus with females living up to two years. This species is found across South America but current stock was collected in Peru in 1997.

Blaptica dubia (Serville, 1838)
Guyana Spotted Cockroach, ACS #13
Adult length: 40-48 mm
Glass climbing: No
Reproduction: Live-bearing
Adult females of *B. dubia* range widely in coloration from nearly solid brown to black, with yellow and orange spots and checkers. The males are slight of build with long tegmina that

Blaptica dubia from the 1990s

Blaptica dubia © Tom Larsen

cover membranous flying wings. Females only have tiny tegmina and no rear wings. Males live less than a year, while females can live two years. Females of the original strain were easily stressed and aborted oothecae consistently without warm temperatures and fruit included in the diet (McMonigle and Willis 2000). The common stock seen today is much more forgiving. Still, they are a little more demanding of high temperatures for breeding than *Blaberus*. Enclosures should be kept close to 80° F (27° C) for good reproduction.

This species hails from Argentina, Brazil, Guyana, Paraguay, and Uruguay. Current stock was kept in the United States at least back to the early 1980s, but it was rare and never used as a feeder. This species was commonly traded in Europe (Mann 1988). It became a popular prey item in Europe by 1991 at the latest. This eventually influenced its feeder status in the states, but not till about 2002. This has become the most familiar and widespread feeder cockroach. Cockroach breeders have worked to develop pale yellow or orange strains from color forms that have shown up occasionally in the original stock (at least for a few decades).

Blatta lateralis (Walker, 1868)
Turkistan Cockroach, ACS #15
Adult length: 22-28 mm
Glass climbing: No
Reproduction: Egg-laying

Male Turkistan roaches are a light yellowish tan with a white margin. They are fully winged. Females have only tiny wings and are russet brown with a few white markings. *Blatta lateralis* can reproduce very quickly in captivity and grow to adults in just a few months. Adults live less than a year with males dying first. Most foods are accepted.

This is one of those fun species when it comes to names. This creature was known as *Blatta lateralis* for a century, until someone noticed a synonym created by an accidental redescription by Adelung (1910) who called his 'new' species *Shelfordella ahngeri*. Someone bastardized that with another accidental synonym by Saussure, 1874, who renamed it *Periplaneta tartara*—so it became traded in Europe as *Shelfordella tartara* for some years. Additionally, Turkistan can be spelled Turkestan. Another common name for this species is red runner, because the nymphs are reddish and they run (like cockroaches).

This species has become widely distributed across the tropics, including warm areas of the Americas, Asia, and Africa. It was introduced to North America by the military—it was first seen in 1978 around a California army base and shortly after at a base in Texas (Mohan 2013). In Arizona they can be very common in water boxes and around irrigation systems around hotels and businesses, but they do not exist in natural areas (pers. obs. 2009). It is common at southern Arizona rest stops from as far West as Yuma to Rio Rico (pers. comm. Clausen). It has become a popular feeder since it does not climb, does not burrow, does not infest homes, is fast, and is easy to breed.

Blatta lateralis female with ootheca © Henry Kohler

Blatta lateralis female with ootheca © Henry Kohler

Blatta lateralis male © Henry Kohler

Blatta orientalis Linnaeus, 1758
Oriental Cockroach, ACS #14
Adult length: 18-29 mm
Glass climbing: No
Reproduction: Egg-laying
This species is widespread and has many names, from croton bug to waterbug (waterbug should not be confused with the predacious Hemipterans called water bugs, which eat other insects and inflict painful bites) but Oriental cockroach is the standard. *Blatta orientalis* tend to live in wet areas, especially in basements around sinks and drains. I remember hunting for these in my grandmother's basement, but all I ever found were a few dead ones. Many years later, as a pest control technician, I would find them in sticky traps in certain department stores—but even the living ones were not salvageable due to the nature of sticky traps and frailty of their exoskeleton. Males are more often found in the traps than females, but oothecae were always dried out or hatched to a rapid, sticky demise. In 1996 I encountered a small captive colony of parthenogenetic specimens an entomologist I worked with had maintained for many years. At the time I was not interested, but a year later I asked about obtaining some. He said they had saved his life and died in the process. His furnace started producing high levels of carbon monoxide; he felt quite tired but did not know why. The sudden death of his *Diploptera punctata* (collected years earlier in Hawaii) and parthenogenetic *B. orientalis* colonies compelled him to search out the problem.

Blatta orientalis from Andalucia, Spain
(CC-BY) Gail Hampshire

The adult female has tiny little tegmina while adult males have both sets of wings. The tegmina extend about two-thirds the length of the male's abdomen but can be much longer. *Blatta orientalis* is one of very few species that can continue on through parthenogenesis over multiple generations in captivity. Feral parthenogenetic populations are currently unknown. Adult females may live up to a year after reaching adulthood, while adult males last considerably less time. These are found worldwide but the discovery of specimens under leaves and stones in the Crimean Peninsula suggest Southern Russia was the original home (Thomson 1921).

Blattella germanica (Linnaeus, 1767)
German Cockroach, ACS #16
Adult length: 10-13 mm
Glass climbing: Yes
Reproduction: Egg carrying
Although a tiny animal, this is the biggest pest of the roaches. Hundreds of German cockroaches can be found under one drawer or behind one picture on the wall of an infested dwelling—and that is dwarfed by the thousands lodged inside the mechanics of a large refrigerator. *Blattella germanica* might make a good feeder roach if it were not such a horrendous pest. A female can produce as many as eight

Blattella asahinai female © Alan Jeon

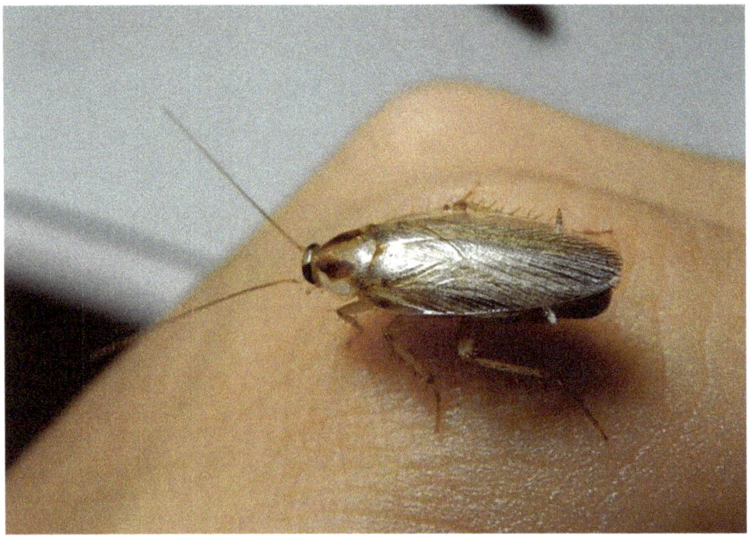

Blattella vaga female © Alan Jeon

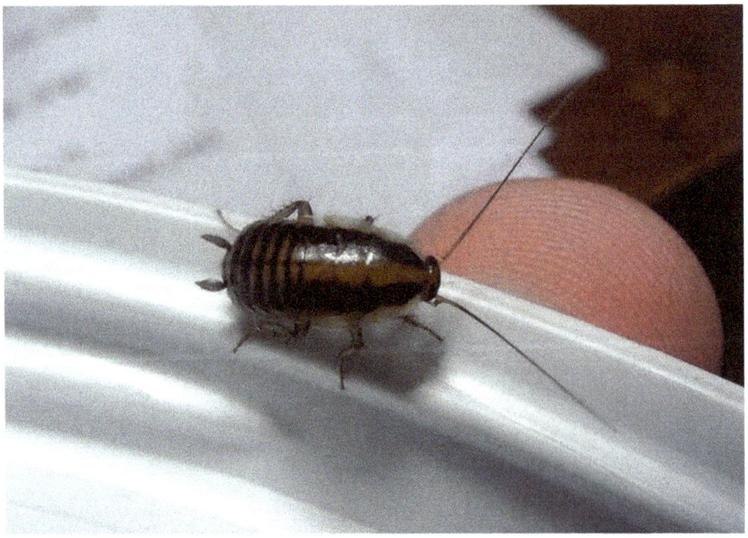

Blattella vaga nymph © Alan Jeon

oothecae in her life. Each ootheca contains up to forty eggs that hatch in about three weeks. The nymphs need another month and a half to reach adulthood under optimal conditions. Adults seldom live over six months. This species is found everywhere man lives. *Blattella asahinai*, a very similar outdoor species native to Japan, India, and tropical Asia showed up in Florida some years back; its arrival made news clips as it was said to dive-bomb people outdoors. It is very difficult to differentiate from *B. germanica* and though it is not know to hybridize in the wild, captive crosses have been studied to understand the relationship between the two (Ross 1993). *Blattella vaga* Hebard, 1935 Field Cockroach ACS #18 hails from India but was originally described from southwestern North America. It is also found outdoors but is easier to tell apart because the stripes on the pronotum are darker and more distinct. *Blattella nipponica* Asahina, 1963 Japanese Field Roach ACS #17 is also closely related, looks similar to the German, and does not reproduce in peoples' homes. Adults fly well. The stripes on the pronotum curve inward at the ends, the opposite of *B. germanica*. *Blattella nipponica* is found living among leaf debris and rotten logs in the forests of its native Japan and is reared by some native blatticulturists. Most of the fifty or so *Blattella* species have probably not been reared in captivity. A black form of *B. germanica* (var. Auburn black German) has been isolated and is reared in some laboratories.

Byrsotria cabrerai Rehn and Hebard, 1927
Striped Burrowing Roach, ACS #19
Adult length: 32-48 mm
Glass climbing: No
Reproduction: Live-bearing
This banded *Byrsotria* is one of the prettiest members of its genus, though just the males have pale bands along the rear margin of the abdominal segments. The female has small yellow triangles running along the edge of the tergites, as do the nymphs. However, the nymphs appear to be uniform in color (unless there is no substrate or a substrate without fine particulates) because the rough surface of their exoskeleton collects a layer of fine dirt. In natural populations, males may have long or short wings, but captive stock males only express short wings (though much longer than the females' tegmina).

This is a very easy, hardy species often used as a feeder. It is much easier to maintain than *B. fumigata* or *B. rothi*. The oldest date for the current culture stock is 1997. Like all *Byrsotria*, it is a Cuban endemic.

Byrsotria fumigata, adult males and female, 1990s stock

Byrsotria fumigata (Guérin-Méneville, 1857)
Cuban Burrowing Roach, ACS #20
Adult length: 32-45 mm
Glass climbing: No
Reproduction: Live-bearing
Male and female Cuban burrowing roaches look very different. The wings of the male are large, while the female's tegmina are short and

Byrsotria cabrerai nymphs are often coated in dirt.

Byrsotria cabrerai adult female (left) and male (right)

Byrsotria fumigata

rectangular like *Hemiblabera* and certain *Eurycotis*, but in this case only the female. Males can have half-length pentagonal tegmina, three-quarter length round-ended wings, or full-length wings resembling the tegmen form of *Blaberus*. Male wing forms tend to be culture specific, but might vary within a population. Nymphs are plain brown with a row of pale spots along the sides of the abdomen. A yellow band runs across the front of the female's pronotum, while the male's is sometimes a very thin band. Males are nearly solid black unlike any other commonly kept *Byrsotria*. Adult males live eight to twelve months while the females can live two years.

This species is found in Cuba. It was cultured in the U.S. at least as far back as 1960 (Barth 1964) and males had long wings. The U.S. stock from the early 1990s was the round-winged form (males), which may date back to the 1960s (Roth and Willis 1960), but it seems to have disappeared by 2005. This species was also listed on the first BCG culture list 1986. However, the UK stock had long-winged males. Current stock may be the same European long-wing form.

Byrsotria rothi Gutiérrez and Linares, 2003
Roth's Burrowing Cockroach, ACS #21
Adult length: 45-58 mm
Glass climbing: No
Reproduction: Live-bearing
This handsome burrowing cockroach is a little larger and more colorful than *B. fumigata* because the colors are more variable. Individuals can be dark brown, almost black like

Byrsotria rothi adult male with long wings © Chen Zha

Byrsotria rothi adult female © Chen Zha

B. fumigata, or have pale tan wings like most *Blaberus*. This species is a little strange because the male wing length is just as variable as color, and not by geographic race. Within a single culture both short-winged and long-winged males and light or dark animals are produced.

Nymphs often take more than a year to reach adulthood, though the males mature faster. Specimens breed slowly and leave each other alone. It can take an incredibly long time for colonies to outgrow their enclosures even with a good diet of dog food and apple. This south Cuban culture stock showed up in captivity around the time of the description.

Corydidarum pygmaea (Karny, 1915)
Pillbug Cockroach, ACS #104
Adult length: 14-18 mm
Glass climbing: Yes
Reproduction: Live-bearing
This insect can roll into a ball when threatened. There are a number of cockroaches that appear similar to common isopods, especially as nymphs, but not many that resemble the volvating isopods known as pillbugs. There are quite a few species, mostly desert cockroaches from the Family Polyphagidae, but also some live-bearing Blaberidae that retract their legs and curl the body into a hemisphere. However, the only truly spectacular examples perfecting this defense hail from the family Blaberidae. There are very few species with this ability and the most familiar is the incredible *Corydidarum (Trichoblatta) pygmaea*, commonly known as the pillbug cockroach. Its common name refers to its resemblance to the volvating terrestrial isopods known as pillbugs. The two types of arthropods look quite similar, as even the volvating plates are similar in design. This can be contrasted against an entirely different volvating creature, the pill millipede, with tapered, recess-margined, non-overlapping segments. In the pillbug and the pillbug cockroach only the thoracic segments overlap, providing a recessed space for the legs, but the cockroach thorax is shorter and smaller in relation to the abdomen. The three thoracic segments of the cockroaches make up only half the sphere, while isopods have seven thoracic segments that comprise three quarters of the sphere. Cockroach antennae are longer, thinner, and comprised entirely of tiny segments, while pillbugs have much thicker antennae comprised mostly of large elbows. Cockroaches only have six legs and a small pair of cerci, while pillbugs have fourteen legs and a multi-segmented tail with a number of small, ventral appendages. Even molting is different, as pillbugs only molt half at a time. The thin white exuvia of pillbugs splits lengthwise between the fourth and fifth thoracic segments, usually with a day between each side. *Corydidarum* molt

Corydidarum pygmaea: A bit of manipulation is required to get a picture of small nymphs since adults always shield them.

Corydidarum pygmaea

Corydidarum pygmaea female, nymphs, and adults

along the midline of the head and thorax all at once, and the exuvia looks like a complete animal. Unlike most pillbugs, it can be difficult to get this pillbug cockroach to roll into a ball, especially for older nymphs and adults. Females are often too full, while males lose the ability entirely at maturity.

Adult *C. pygmaea* males look like a normal roach and cannot curl up into a ball. Males and females appear similar when small, but by subadulthood the male has large wingpads. At adulthood the two genders look like entirely different animals. Despite great physical differences, the male's most unusual characteristic is his miniature size compared to the female. She is not a big cockroach herself, but matures slightly larger than the largest specimen of the largest species of the familiar pillbugs. He is nearly as long but hardly a quarter of her mass. He has long wings and a sculptured pronotum with a shape that gives the overall impression of a miniature polyphagid (this fact was not lost on the original describer, as *Corydia* was a polyphagid genus). Another similarity to some of the desert roaches is the relative longevity. The male rarely survives two months, while the female lives one to two years. She only needs to mate once and can continue to reproduce for a year or more. Her fertility and reproductive readiness are obscured by body structure so much that it is very difficult to tell a young adult female from one ready to give birth. Two to three times a year she can produce one to two dozen nymphs. Females often will not give birth until six months to a year after the males have died. They commonly shelter the young under their bodies for at least the first few molts.

This uncommonly encountered cockroach from Taiwan and Okinawa, Japan, was rumored to feed on fungus in nature (McMonigle and Willis 2000), but common food molds and commercially available mushrooms are usually avoided like the plague. In captivity the favorite food is tiny apple slices, but dry dog food is also eaten. Most other fruits and vegetables are minimally, if at all, fed upon. With a regular feeding schedule this species has proven easily maintained, but small nymphs cannot be neglected (McMonigle 2013b). Even with excellent care they grow slowly and are not very productive. Pillbug cockroaches reproduce and grow far too slowly to be used as prey for other invertebrates.

The only drawback to keeping these, and another anatomical difference from pillbugs, is this cockroach has adhesive pads on the feet. It is an excellent climber of smooth and even lubricated surfaces. Tight, escape-proof caging is a necessity. The small nymphs are especially good at climbing and are difficult to contain when using a petroleum jelly barrier. Sometimes 1st instars will get stuck backwards on the petroleum and if not found and removed within a day they will desiccate and die. Small, thin pieces of curved bark laid on the substrate will give all stages a hiding spot and can help reduce climbing. Too much or too little moisture will increase climbing behavior. A tight fitting lid is helpful, but early instar animals hide in the seams and are easily flung from the cage or smashed when the lid is opened and closed for maintenance. A liquid Teflon barrier works spectacularly to contain this species.

Recently someone noticed the genus was first described as *Corydidarum* 1865, and *Trichoblatta* 1895 later. Nearly all scientific and popular references to this species refers to it as *Trichoblatta*. Japanese specimens of this tropical Asian species were featured in the *Allpet Roaches* (2000) book. At the time it was doubtful livestock would ever become available. There was no reason to believe this dream species could even be kept in captivity successfully. Current stock may only date back to 2010.

Corydidarum pygmaea group with male (rusty individual recently molted)

Corydidarum pygmaea group

This genus contains an even more spectacular species, *Corydidarum magnifica* (Shelford, 1907) from China, which looks similar but is larger, metallic-green, and not yet in culture.

Cryptocercus punctulatus Scudder, 1862
Brown Hooded Cockroach, ACS #22
Adult length: 24-30 mm
Glass climbing: No
Reproduction: Egg-laying

Although there are many native North American "wood roaches" (*Parcoblatta* and others), *Cryptocercus* are the only ones that feed solely on wood. Small nymphs look a lot like termites and stay with the parents until maturity. At each molt they shed their stomach lining and need to replenish the flagellates by ingesting the parents' frass. Without them, the roaches slowly starve to death. They live in small colonies commonly composed of an adult pair and a number of nymphs, however, larger, more diverse groups can be encountered in large rotten logs. They are the closest living relatives of termites (termites are presently considered eusocial cockroaches).

Cryptocerus clevelandi found in a fallen log, Cave Junction, Oregon © Peter Clausen

Cryptocercus punctulatus from Mercer County, West Virginia: usually a pair and some nymphs are found in galleries within damp rotten logs laying on the forest floor, usually with a diameter from 5″- 20″(13-51 cm). The logs chosen are very rotten and can usually be broken apart with bare hands.

The brown hooded cockroach is a uniquely fascinating cockroach in relation to its habits and it is also a relatively large, handsome species. The adult is wingless, red-brown to nearly black, and both sexes have a pair of large protrusions on the pronotum somewhat like *Elliptorhina*. Since they are large, wingless, and superficially resemble the better-known hissing cockroaches, they are the cause of the occasional North American reports of "hissers out in the wood pile." Nymphs start out with a see-through exoskeleton, resembling giant termites until they have molted a few times.

Cryptocercus are difficult to keep in captivity because they can die from handling and grow very slowly. Nymphs require up to five years to reach maturity, with less than four years highly unlikely. Young can survive without the parents, but do not seem to grow as large (McMonigle 2009c). They prefer to live inside rotten logs and tend to waste away if not provided this source of food and shelter.

Smaller chunks of rotten wood can be used as food, but the animals tend to die within a few months without the shelter provided by a log (minimum 18″ long by 4″ diameter). Lastly, as with many termites, if they get too warm the flagellates in the gut can be killed and the animals will starve. Temperatures much above 72° F (>22° C) can kill the symbiotic flagellates.

The genus *Cryptocercus* is normally placed in its own family, the Cryptocercidae. Until recently it was believed there was only a single North American species, *C. punctulatus*, and two Chinese species. However, DNA analysis from 1996 to present has led to three additional species descriptions in the eastern U.S. and one in the Northwest. The result is five North American species that are physically identical and likely have no specific barriers to mating. For this reason, outside of DNA analysis, the only way to obtain a tentative identification for North American species is by the collection location. On the other hand, Chinese species have expanded to five as a result of female genitalia diagnosis (Wang et al. 2015). The *Cryptocercus punctulatus* specimens pictured here are from Virginia, but all the species in this genus fall approximately into the same range of color, size, and shape.

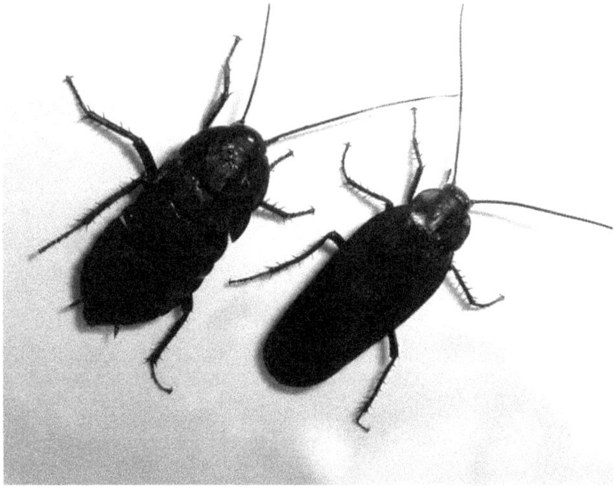

Deropeltis erythrocephala adult pair

Deropeltis erythrocephala (Fabricius, 1781)
Black Velvet Cockroach, ACS #23
Adult length: 38-49 mm
Glass climbing: Yes
Reproduction: Egg-laying

The black velvet roach is a beautiful midnight-black cockroach with a satiny finish. It was the largest available egg-laying species for the dozen or so years it was kept in culture (1997-2009) but it appears to have been lost from culture. The oothecae are huge (25 mm), black, and look like an *E. floridana* ootheca if it were half again longer. Nymphs are dark brown with reddish-orange legs and an orange-red head. The adult male flies well and has long wings that resemble black tissue paper. The big velvety female has no sign of wings and is often the longer of the two genders. Adults live up to a year and are sensitive to lapses in care, especially dryness. This species occurs in Namibia, South Africa, and Zimbabwe.

Deropeltis sp. 'Jinka' is a more recent species whose adults look similar, though shorter (28-34 mm), and only a fraction of the mass. The nymphs have dull, brownish-orange heads and legs, unlike the brightly colored *D. erythrocephala*. This stock comes from Ethiopia.

The genus *Deropeltis* contains nearly fifty large, egg-laying species found across the African continent. L. M. Roth was working on this genus and had me send specimens of cultured species in 2001-2002. Unfortunately his work went unfinished. From 1994 to present there have been as many as six different bloodlines of *Deropeltis* species that looked relatively similar, but perhaps were just two or three with alternate names. Kevan and Chopard (1954) reported some *Deropeltis* species from dry grasslands in Kenya and Somalia, but current

Black cockroach, *Deropeltis* sp., Kruger National Park, South Africa (CC-SA) Bernard Dupont

Deropeltis paulinoi small nymphs

Deropeltis paulinoi male

Deropeltis paulinoi female

Diploptera punctata

species stop growing as nymphs and do very poorly in dry enclosures, even if moisture is available from a water dish.

Deropeltis paulinoi oothecae

Deropeltis paulinoi Bolívar, 1881
Ornate Velvet Cockroach, ACS #24
Adult length: 28-38 mm
Glass climbing: Yes
Reproduction: Egg-laying
This species is found in Angola, Botswana, Congo, Namibia, South Africa, and Zambia.

Males are slight with large wings, while females are physically similar to large nymphs. Nymphs are normally black with orange markings or legs, while adults tend to be solid black or brown. *Deropeltis paulinoi* is a medium-sized species and the lighter-bodied males tend to be the longer of the two genders due to their wings. Adult females of this species retain the coloration of late instar nymphs, which are predominantly black with orange markings. (Early instars have a reverse pattern, mostly orange with a large dark spot.) Males are jet black like other *Deropeltis*.

Females produce dozens of large oothecae, which are often glued under objects. They should be misted occasionally and kept at 75° F (24° C) or higher, to hatch after three to four months (McMonigle 2007b). Although oothecae do not hatch with total neglect, they are far sturdier than other *Deropeltis* and hardier than most egg-laying species.

Deropeltis paulinoi is an extremely productive species, but it probably would not be eaten by the majority of insectivorous vertebrates. The orange-and-black nymphs are warningly colored for a reason. *Deropeltis* emit quinones as a defensive secretion. The secretion is similar to quinones produced by giant round millipedes. The pungent chemical smell is noticeable, but the quantity of defensive fluid is minimal compared to giant millipedes. Although nymphs make a successful food for water-scorpions, mantids, and tarantulas, it is important to use forceps during feeding. Forceps are clumsy, but they are needed to prevent fingertips from turning brown for the next few days. Also, even arachnids have to be relatively starving to eat these. They may be blind, but they are not stupid.

Unique and superb color along with appreciable size ought to make this a highly sought-after species. However, reproductive ease and

Deropeltis paulinoi female and male subadult

ready acceptance of overcrowded conditions should make it a low-priced feeder. On the other hand, quinone production greatly detracts from its value as a feeder. Nevertheless, the ornate velvet cockroach is one of the sharpest looking species to enter the hobby in the last ten years.

Diploptera punctata (Eschscholtz, 1822)
Pacific Beetle Mimic Cockroach, ACS #25
Adult length: 18-25 mm
Glass climbing: Yes
Reproduction: Live birth

Despite a muted and unassuming appearance, *D. punctata* is one of the most fascinating and unique cockroaches in the world. Physical and olfactory mimicry and reproductive strategy set apart this handsome species. It is a native of Asia but has been spread around the tropics by man. In the U.S., it is widespread in Hawaii. *Diploptera* is the only genus in its subfamily. There are six other members of the genus *Diploptera*, but none of them have been kept in the hobby.

Visual mimicry is heightened by olfactory imitation. As the common name suggests, the adults look very much like a beetle. The wings are formed in such a way as to suggest a straight line down the middle where a beetle's elytra meet, rather than the overlapping tegmina of the common cockroach. The model is probably a medium-size, dark-brown tenebrionid (darkling beetle), but there is some resemblance to dytiscids (water beetles). This cockroach is capable of discharging a defensive secretion that smells (and probably tastes) the same as the common darkling beetle defensive secretion. It does not smell the same as quinone secretions of spirobolid millipedes or *Deropeltis* cockroaches.

Diploptera punctata's defensive secretion is of similar chemical composition, but is minimally potent compared to some darklings. The secretion is ejected from the second pair of abdominal spiracles and is created by glands surrounding the tracheae. It is composed of *p*-benzoquinone and two of its derivatives (Eisner 1958). Darklings produce a mix of benzo-, ethyl-, and methyl-quinones from glands near the end of the abdomen.

Wyttenbach and Eisner suggest that *Diploptera punctata* is unusual among roaches because the female mates immediately after a molt, before her exoskeleton hardens (Wyttenbach and Eisner 2001). Her secretory glands do not function for the first few hours and they suggest the male offers some protection, although he does not discharge the defensive secretion during mating if she is attacked, only when he is himself disturbed. It sounds like a reasonable theory, but the majority of roach keepers are aware it lacks a key detail. Mating with freshly molted females is a common behavior for most, if not all, of the blaberid species kept in captivity, many of which have relatively impotent defensive smells or fluids. The behavior is a result of mate competition, has nothing to do with defensive glands, and is more likely to lead to the death of a teneral female than salvation (McMonigle 2010a, 2012d).

What sets this species apart from most invertebrates, and makes it a singular species among the cockroaches, are the specifics of its reproductive methodology. It is the only known viviparous cockroach. That is, the female nourishes underdeveloped eggs internally until they are ready to be born. The large relative size and low number of nymphs per brood are byproducts of this strategy.

The Pacific beetle mimic is easy to keep in captivity and has been in culture for a long time (Roth and Willis 1955). There have been multiple sources of stock over the years. The few

Diploptera punctata nymphs and adult

Dorylaea orini nymph

stocks with origin information were collected in Hawaii. It reproduces when kept as low as 70° F (21° C), but is more productive at 77-86° F (25-30° C). Despite small broods, it can reproduce quickly enough to be used as a feeder. Moisture should be provided in the form of a water dish or damp areas of the substrate. General fare of dog food is useful but fruit can also be added.

Dorylaea orini adult

Dorylaea orini Roth, 2002
Orin's Cockroach, ACS #26
Adult length: 27-34 mm
Glass climbing: Yes
Reproduction: Egg-laying

Dorylaea orini nymphs start out brown, but become strikingly marked in black-and-white banding after a few molts. Nymphs are highly attractive and rival any other species. They are among the most visually striking roach nymphs. Adults have long, dark-brown wings and are not among the more attractive roaches. Nymphs and adults can climb glass, but are easily kept in check by a petroleum jelly barrier. Females produce dark brown oothecae containing approximately thirty eggs. The egg cases are glued lengthwise to rotten logs, while tiny bits of chewed wood are carefully glued over the exposed side. If there are no good spots for laying, the cases will be dropped, left exposed on the substrate. The female may then still glue materials to the surface of the ootheca.

Although it hails from tropical Malaysia, specimens do well at room temperature around 72° F (22° C). They may grow more rapidly at higher temperatures but are more prone to desiccation. Some dryness is acceptable, though moisture should be available for drinking at all times and oothecae need to be misted regularly unless the humidity is above 75%. Good ventilation is required—microscreen should be used for vent covers to help maintain humidity if ambient humidity is low. The roaches do well on a diet of dog food and will also feed on many fruits and vegetables. Adults survive about eight months.

This is an easily maintained, visually appealing species that quickly became one of the more popularly kept roaches. It is a good feeder for amblypygid immatures, but it is kept almost entirely for observation.

This beautiful species was described in 2002 by Dr. Luis M. Roth of Harvard University, but entered the hobby a few years before it had a name. Originally it was labeled as an unknown species of *Periplaneta*. It can be difficult or impossible to find scientific names for unidentified roaches. It is extremely rare for a species not to have a name, but it is common to not be able to determine the name. Unless the animal is a well-publicized species or has uniquely obvious specific characteristics, the only method is to send male and female specimens to a taxonomist. The male specimen is most important since its dissected genitalia are a key indicator. Identification is not the easiest task since there are very few taxonomists

Dorylaea orini nymphs

and each specializes in specific groups (as often determined by geography as systematics). Even if a taxonomist can be found for the type of roach to be identified, the roach may not be identifiable without exact collection location information. It may not be an easily identified species even using dissected genitalia, or the information may not be readily available. Most important of all, the taxonomist must be willing to offer his or her time.

Simply running across an undescribed species does not mean much. I have sent many cockroach specimens to various taxonomists that probably will never be examined (though only after first asking if it is OK to send and for their preference on how specimens are preserved). Some have been identified, but many more have sat for years on shelves and possibly made their way to circular files by now. With *Dorylaea orini* I was lucky. My hope was to find the correct name, but to have discovered a new cockroach was both a surprise and a thrill. Dr. Roth graciously offered to take a look at a variety of dead specimens of different stocks for identification. He determined this one was a *Dorylaea* species and checked for previous descriptions. Later he said he would formally describe this new species and kindly asked what name I would like. I asked him to name it 'pulchra' (meaning beauty since the nymphs are very pretty) but he responded, "I wouldn't name

Drymaplaneta semivitta, Gisborne cockroach © Tibor Ludányi

Drymaplaneta semivitta, Gisborne cockroach (oothecae are pointed at one end) © Tibor Ludányi

Drymaplaneta semivitta, Gisborne cockroaches mating © Tibor Ludányi

it 'pulchra.' There are already eight species of cockroaches named 'pulchra' in other genera. There are lots of other words for beautiful if that's what one wants to name it." Then he asked if I wanted to use my last or first name for the species. (I chose my first name but may have decided differently if my name were John, Mike or Bob.)

Dr. Roth, who described this species and who was arguably the most influential cockroach taxonomist in the 20th century, passed away in 2003. Roth told me he published the description as part of a larger revision of Asian blattids. I had no information where he published the description or what happened to the holotype and specimen series. He said, "The Dorylaea is part of a paper that includes a new genus" (Roth pers. comm. 2002), but that cryptic explanation was the last I heard. In 2012 I encountered the listing for the holotype specimen deposited at Harvard in 2002. The holotype is an adult male specimen listed as *Dorylaea orini* under Blattodea: Blattidae and the type number is MCZT_35085 (MCZ type database @ Harvard Entomology). Modern taxonomists are constrained by ICZN rules that require a type specimen be deposited at a museum for any description to be valid (a few decades ago there were no constraints, which one reason many species described in the past cannot be validated and have no type specimens). One of the first articles on this species (McMonigle 2006b) contained photographs and details of the species, though it listed the description as 2003. However, the date was 2002 per Roth's holotype label or, less likely, validated at the time of the (McMonigle and Willis 2002) Allpet Roaches handbook ISBN version, containing the new species name, a short description, and photographs of nymphs and adults (McMonigle 2011c).

There are approximately 25 *Dorylaea* species found in southern Asia and the Indopacific. *Dorylaea orini* hails from tropical Malaysia. *Dorylaea magna* (Shelford, 1909) is the only other species that has been seen in culture, but likely not past one generation. It is a large (~45 mm) reddish-black creature from Borneo.

Drymaplaneta semivitta (Walker, 1868)
Gisborne Cockroach, ACS #103
Adult length: 24-30 mm
Glass climbing: Yes
Reproduction: Egg-laying
This species gets its common name from the city in New Zealand where adventive specimens were first discovered. It was endemic to Australia. Stocks have come directly from Australia and were called Gisborne cockroaches at the time (Mann, 1988), or have come from New Zealand (Ludányi 2012b). It is a handsome species that reminds one of *Eurycotis*. However, it constucts a strange ootheca with a pointed end and it may be a beneficial species. Reports suggest this genus is helpful to man because it hunts other insects in shrubs and trees (Roth and Willis 1960). Although the Gisborne cockroach has been maintained off and on over the years, and is also kept in Japan, it is a rare culture item. Another species, the common shining cockroach (*Drymaplaneta communis*), which lives in Australia around Melbourne and Sydney, has been in the news lately as it has become more common around homes.

Elliptorhina chopardi (Lefeuvre, 1966)
Dwarf Madagascan Hissing Cockroach, ACS #29
Adult length: 36-42 mm
Glass climbing: Yes
Reproduction: Live-bearing
Elliptorhina chopardi is a small, tan and orange species and is still the smallest hisser maintained in captive culture. The dwarf Madagascan hisses loudly when disturbed like its larger relatives. The genus name *Elliptorhina* translates as 'elliptical horn.' *Elliptorhina chopardi*'s horns are very small and run at opposite angles

Elliptorhina chopardi male

Elliptorhina chopardi

Elliptorhina javanica

along the pronotum in an ellipse or V shape. In contrast, the horns of *Gromphadorhina* species start near each side and go forward in a U shape.

The species does not enjoy eating green vegetables like its larger relatives and usually refuses leaves. It prefers dog food and fruit slices. *Elliptorhina chopardi* is relatively productive and is sometimes used as a feeder. Adults rarely live past two years. This is the oldest hisser stock other than *G. portentosa*. Current stock dates back at least to the mid-1980s.

Elliptorhina davidi Herrewege 1973 Bumpy Hissing Cockroach ACS #110 looks a little like a large, up to 55 mm, version of *E. chopardi,* but with large bumps or knobs all over the pronotum and body. However, some specimens lack the bumps. These began to show up around 2013 and are always listed as "sold out" on vendor lists. The adults are similar in color to standard hissers; orange with a black pronotum.

Elliptorhina javanica (Hanitsch, 1930)
Javan Beauty Hissing Cockroach, ACS #93
Adult length: 36-46 mm
Glass climbing: Yes
Reproduction: Live-bearing
This beauty (actually from Madagascar) averages maybe 20% larger than *E. chopardi* and is more contrastingly marked. The body segments are cream-colored, followed by a yellow

Elliptorhina javanica male and females

Elliptorhina laevigata male adult

Elliptorhina laevigata adult pair

to brown band and then a black margin. The adult often has large white markings on the pronotum, moreso than *E. chopardi*. It is longer lived, often reaching two years as an adult, and it is far more difficult to get young from. A heat mat or temperatures around 80° F (27° C) may be needed if females refuse to produce young. These colorful hissers showed up in culture around the year 2000. Some vendors recently changed the name to Halloween hisser, but *E. laevigata* and *G. portentosa* have also been sold by that name since they are actually black and orange.

Elliptorhina laevigata (Saus. and Zehnt. 1895)
V-Horn Hissing Cockroach, ACS #30
Adult length: 44-77 mm
Glass climbing: Yes
Reproduction: Live-bearing
This colorful hissing cockroach is a hobbyist favorite. V-horns survive and grow under most conditions, but require warm temperatures (77-86° F; 25-30° C) for proper gestation. This is a large, elongate species with huge, sausage-like females. The basic horn structure is the same as for the smaller *Elliptorhina* species. The mesonotum, metanotum, and abdominal segments are black, usually with an orange margin at the front of each. Females are often close to 3″ (75 mm) in total length, while males are more variable and range from 45-65 mm. Large males have more massive heads, legs, and segments than females, but they are shorter primarily because their abdomen is not distended. Adults live about three years. Current stock dates back to specimens imported from Madagascar in 1997.

Epilampra maya Rehn, 1903
Mayan Roach, ACS #132
Adult length: ~30 mm
Glass climbing: Yes
Reproduction: Live-bearing
The subfamily Epilamprinae, named for this genus, includes all of the known aquatic cockroaches (Shelford 1907). These are moderately large species that live among rocks and leaf debris along stream and riverbanks. They dive or run into the water when predators and roach collectors approach. *Opisthoplatia* and most *Rhabdoblatta* are in the same subfamily but are less publicized for their semi-aquatic behavior. *Stictolampra lurida*, the most promoted captive aquatic species, was collected from the banks of the Sarawak River in Borneo in 1994 (Bragg 1997b) and was kept in captivity, but it died out after a few generations (McMonigle and Willis 2000). Another semi-aquatic relative collected from the same river banks was *Rhabdoblatta circumdata*, but it never made it through a generation since just one male nymph made it back alive (Bragg 1997b). The cover for Bragg's cockroach rearing manual is a line drawing of the famed *S. lurida*. Superficially it looks similar to *E. maya*.

The small, handsome *E. maya* is found in Central America, Mexico, and Florida, USA. This species has been collected in Florida on a regular basis, but it is not easy to find unless the collector knows the right streams to flip rocks by. This aquatic species may prove difficult to maintain over the long run like its similar predecessor (*S. lurida*).

Eremoblatta subdiaphana (Scudder, 1902)
Hairy Desert Cockroach, ACS #31
Adult length: 9-12 mm
Glass climbing: No
Reproduction: Egg-laying
The hairy desert cockroach lives in the sand in its native range, but is found mostly where plant refuse accumulates, especially piles of tiny mesquite leaves. It has a wide range, all the way from Texas to California, but it seems

Epilampra maya adult © Alan Jeon

Epilampra maya nymphs © Alan Jeon

to be far more difficult to find than *Arenivaga*. In California the native polyphagids are restricted to the Central Valley, deserts, and southern coastal dunes. Here they are evidenced by serpentine surface ridges on the sand (Powell and Hogue 1979).

A habitat terrarium may allow the energetic burrowing habits to be viewed if set up like an ant viewer. Mostly only the adult males end up being seen on the surface, for the short time they live. Males are fully winged, pale, and a little fuzzier than the average polyphagid male. Females are wingless, round, notably hairy, and subterranean like the nymphs. The leathery oothecae take at least a few months to hatch, but it is difficult to pin down because the few oothecae collected from the female's rump and isolated from the main enclosure failed to hatch. Old, hatched oothecae may be nibbled on, but healthy oothecae are safe from cannibalism. *Eremoblatta* grows slowly for such a small creature and nymphs take up to two years to reach adulthood. Adult males live only two months, although adult females live two years or more. Captive stock of this species lasted four or five years in the late 1990s and there are probably no currently viable stocks.

Ergaula capucina males

Eremoblatta subdiaphana nymphs

Ergaula capucina (Brunner, 1893)
Burmese Beetle Mimic, ACS #23
Adult length: 25-30 mm
Glass climbing: adult males
Reproduction: Egg-laying

This handsome species remains among the oldest polyphagid stocks kept in culture and the first widely kept representative with winged females. Stock origin is listed as Mount Mooleyit, 1200-1400 M altitude in Burma (Durkin 1996) but it is also known from Thailand. Adult *E. capucina* males are standard form for this group with tapering, elongate wings, and a slight build. They are brown overall with a variable stripe across the middle of the tegmina. Adult females are much bulkier, round, have no lighter markings, and the wings fold together to suggest the suture of beetle

Ergaula pilosa adult pair and large nymph

elytra (outer wings). The adult female may mimic the appearance of carrion beetles (Coleoptera: Silphidae). The cockroaches have eversible odor glands along the sides of the abdomen that can produce a rotting flesh smell like the odor produced by some *Silpha*.

After the ultimate molt, females live about nine months, while males rarely live more than three. The leathery oothecae have more than a dozen longitudinal ridges on each side. Around eighteen nymphs hatch after three months. Nymphs take eight or so months to reach adulthood but can take much longer if kept cooler than 72° F (22° C) or fed only vegetable matter.

Ergaula pilosa (Walker, 1868)
Black Beetle Mimic, ACS #107
Adult length: 27-35 mm
Glass climbing: adult males
Reproduction: Egg-laying

This nice-looking species is one of the newest roaches propagated in captivity. Although it has been sold as the "big black beetle mimic," the tegmina color is brown when viewed from most angles. Although the specimens in the accompanying photographs appear to look black, by eye it is brown. The silphid-like female appears to be reddish-brown, rather than black, because of a layer of reflective setae (hairs) on the surface of the wings. The species

Ergaula capucina pair and nymph © Peter Clausen

Eublaberus distanti

Ergaula pilosa female

name *pilosa* means hairy and certainly refers to the dense mat of golden brown setae covering the wings.

According to Beccaloni's Species File, *E. pilosa* is found in Indonesia (Java Island and Sumatra), Malaysia (Malacca State), and Borneo. *Ergaula capucina* looks similar, but is smaller and the males normally have a complete and much wider light band across the wings. *Ergaula capensis, E. carunculigera, E. funebris, E. nepalensis,* and *E. silphoides* are the remaining members of this genus, though none of these are currently known to the hobby.

Rearing appears to be rather straightforward and the life cycle standard for a polyphagid. Oothecae are buried in the substrate and hatch after four to five months. This is a tropical species, so substrate should be kept somewhat damp, not soggy, for development of both animals and oothecae. *Ergaula pilosa* nymphs take a year or more to mature when fed regularly on dead leaves, small amounts of dog food, and fruit. Temperature does not seem to have a notable effect unless they are kept too cool, which could result in slower growth or imperfect molts. The rather hairy nymphs are normally provided an inch or two of substrate to burrow into and food is placed on top. Uneaten dog food and fruit should be removed after a few days. Once they reach adulthood the females can live a year. Males seldom last more than a few months (McMonigle 2014d). With time it is a very prolific species.

Ergaula pilosa is fairly new to blatticulture (2012?) and is certainly easy to propagate. It is a handsome species and among the largest polyphagids.

Eublaberus distanti nymph

Eublaberus distanti (Kirby, 1903)
Six Spotted Roach, ACS #33
Adult length 44-56 mm
Glass climbing: No
Reproduction: Live-bearing
The six spotted cockroach is not a bad wing biter, nor predatory like its familiar relative *E. posticus*. Nymphs are shiny dark brown with six tan-yellow spots (three on each side). Six spots are not definitive, since adults have none, subadults may have only four, and very small or unusual nymphs lack spots altogether. About eight months is taken from birth to adulthood. This species is found from Guatemala to Brazil—the original stock of today's creatures was collected from Trinidad in the 1980s.

Eublaberus posticus

Eublaberus serranus, Pantanal cockroach, adult with headlamp pattern on pronotum

Eublaberus serranus nymph

Eublaberus serranus, Pantanal cockroach

Eublaberus posticus (Erichson, 1848)
Orange Head Cockroach, ACS #34
Adult length: 38-48 mm
Glass climbing: No
Reproduction: Live-bearing
The orange head is among the most colorful of the large roaches and even the nymphs are a beautiful maroon color. The heads are black—the common name comes from the background color of the pronotum. The black markings on the pronotum are variable but often immortalize Groucho Marx.

These are notorious wing biters and they will eat the entire wings off one another if prey is limited. When nearly any type of slow-moving, thin-shelled, or newly molted arthropod (walkingstick, grasshopper, beetle larva, etc.) is tossed in the cage of a colony, the orange heads eat it alive. Maggots are usually eaten first, sometimes only, if infested fruit is introduced. While they exhibit predatory behavior in captivity moreso than any other cockroach, they may not be active predators in nature. They are one of a handful of large species associated with bat caves, where they live among the guano and feed on fallen debris and, less often, fallen bats. Despite a love for invertebrate prey, this species is less likely to cannibalize under adverse conditions than most.

Eublaberus posticus hails from northern South America and southern Central America.

Eublaberus sp. "ivory" adult pair and nymph © Peter Clausen

Current stock dates back at least to the mid-1980s and has no collection data. One specimen was found in imports to the southern U.S. (Hughes 1949). For a long time, this species, *G. portentosa*, and *B. discoidalis* were the only livestock other than pests or small natives that could be found; it is still one of the more widespread stocks.

Eublaberus serranus Lopes, Oliveira and Khouri, 2015
Pantanal Cockroach, ACS #106
Adult length: 36-47 mm
Glass climbing: No
Reproduction: Live-bearing

The adults are narrow in form and a bit less massive than other commonly kept members of the genus, but are still among the larger live-bearing cockroach species. Like some other tropical New World Blaberinae (notably *Lucihormetica* and *Hormetica*) there can be a pair of large white dots located on the dark colored pronotum. These markings are a little different because they are situated towards the rear sides of the dark central spot. The markings are unrelated to sexual dimorphism in this species and are closer to the edges, prompting the name 'headlamp cockroach.' The spots are inconsistent and often look like dents in the side of the dark splotch rather than spots.

Sometimes, not a single adult in a large group has the markings.

The solid, dark-brown nymphs are substrate swimmers and can be kept in very high densities with a few inches of soil and adequate food. Eight to ten months are required for them to reach maturity. Adults can burrow but they prefer to hide under objects or between the slats of roach huts if available. A decent diet and temperatures above 73° F (23° C) are needed for females to produce young (McMonigle 2014b). Wing-biting behavior is common in heavily populated cages, but usually only damages the last 10% of the wing.

The Pantanal cockroach is a handsome, hardy species sometimes used as a feeder, but more often as a pet. Though it has been around since at least 2009, it is extremely new relative to the other *Eublaberus* species maintained in captivity. This species was described from specimens collected in São Paulo, Brazil. Current stock is named after the Pantanal region of southern Brazil where it was collected. A key difference between *Blaberus* and *Eublaberus* is that the adult's eyes are very large, with only a small space between them at the top of the head on *Blaberus*. There is a far larger space between the eyes for *Eublaberus* species, though the difference is not as pronounced in *E. serranus*.

Eublaberus sp. 'Ivory'
Ivoryhead Cockroach, ACS #111
Adult length: 45-58 mm
Glass climbing: No
Reproduction: Live-bearing
These were the feeder roaches cultured in massive 40 ft. x 10 ft. (12.2 m x 3 m) vats at Agama International (a famous reptile breeding center) for decades. (The vats were incorrectly reported to contain *B. discoidalis* in Allpet Roaches 2000.) The late Bert Langerwerf acquired them from South America sometime in the 1980s. This species is responsible for aiding some historic reptile breedings. For a while it was traded as *Eublaberus distanti*, and it is very similar in size and coloration. Beyond moderate differences in the male genitalia, the background color of the adults' pronotum is almost white (ivory). This is a contrast to the golden brown background of *E. distanti*. The dark markings on the pronota are variable by individual, but do not exactly match the pattern of any individual *E. distanti*, which are also variable. Ivoryhead cockroaches can be very productive, but still take about six months to reach maturity, so it takes time to build a large colony from a couple dozen or fewer individuals.

Eucorydia aenea dasytoides nymphs © Chen Zha

Eucorydia aenea dasytoides (Walker 1868)
Taiwanese Flower Cockroach, ACS #127
Glass climbing: Adults
Reproduction: Egg-laying
The metallic beauty and bright colors of the adults are rivaled by few species. Both genders have a similar pattern of colors but they are minimally sexually dimorphic. The males have long wings that cover the abdomen; the tegmina of females are a little shorter and the edges of the abdomen stick out on either side.

Eucorydia aenea dasytoides male © Chen Zha

Eucorydia aenea dasytoides female © Chen Zha

Eucorydia aenea dasytoides nymphs and oothecae © Chen Zha

Of course the appeal of this species is the metallic-blue-and-green coloration and bright orange banding of the adults. The beautiful adults live just a few months. The unattractive, hairy little nymphs live much longer. They are tan and only easy to tell from the average polyphagid nymph because of a white band near the end of each antenna. The tiny serrated oothecae contain a dozen or fewer eggs. The eggs of *Eucorydia* develop quickly for a polyphagid, just one to two months. The time required to reach maturity can be as short as 300 days at 80°F (27°C) (Zha 2015). On the other hand, the Japanese *Eucorydia yasumatsui* take an average of 625 days for females and 608 days for males to develop from hatchling to adult at 64-75°F (18-24°C) (Fujita et al. 2011). Viable multi-generational colonies are likely possible, but *Eucorydia* species were kept little, if at all, prior to 2009. Current stocks are from Taiwan, but it can also be found in China and Vietnam.

Eupolyphaga sinensis (Walker, 1868)
Chinese Medicinal Cockroach, ACS #112
Adult length: 28-34 mm
Glass climbing: No

Eupolyphaga sinensis oothecae, varying sizes

Reproduction: Egg-laying
This animal was used exclusively in Chinese medicine for hundreds of years, though increased demand over the last decade has resulted in farmers switching increasingly to *Periplaneta*. There were a slew of popular news stories in 2013 about 1.5 million *Periplaneta* escaping one of these New Age farms in China. Pastes made from ground-up animals are said to cure a variety of ailments, but folk remedies have limited scientific track records. Nevertheless, an extract made from *Eupolyphaga sinensis* possesses tumor-inhibiting qualities (Ge et al. 2012). This species comes from western China and Mongolia. This is a far larger animal than I expected it to be, and the female is unusually tall for a polyphagid. Eggs hatch after about three months from small oothecae that usually contain fewer than ten eggs.

Euthyrrhapha pacifica (Coquebert, 1804)
Pacific Two-spot Cockroach, ACS #139
Adult length: 8-9 mm
Glass climbing: Yes
Reproduction: Egg-laying

Eupolyphaga sinensis pair, note the male`s big bug eyes

Euthyrrhapha pacifica in Hawaii
(three views) © Cory Campora

Eurycotis decipiens

Although a tiny creature, *E. pacifica* is a fantastic little animal with unique coloration and very strange wings. The tegmina are rounded like beetle mimic species, but they are stuck in a half open position and the flying wings lay flat and unfolded. They look like a lady beetle getting ready to fly or that had a bad molt. This species is common across Africa, including Tanzania, but has also been introduced to most tropical areas accidentally by man. It is common in Hawaii and it seems probable it has found its way to the continental U.S.. The oothecae have crackled skin somewhat like lizard skin. This species has little or no captive history, but it is included here because it is surprising stock has not been established already.

Eurycotis decipiens (Kirby, 1903)
Costa Rican Zebra Cockroach, ACS #35
Adult length: 28-36 mm
Glass climbing: Yes
Reproduction: Egg-laying

The zebra cockroach is striking as both nymph and adult. *Eurycotis decipiens* is one of the older pet species, but remains among the more 'collectable' and desired. The adult's dazzling coloration and outgoing behavior are major reasons for its favor. It has also proven a challenging species for many hobbyists to keep through multiple generations, which keeps its numbers in check. The currently traded cultured stock dates back at least to the early 1980s and is rumored to have been collected in Costa Rica. However, it may be the same stock listed as the Trinidad zebra on the Cincinnati Zoo Animal Inventory List, 1993. The species was described from Trinidad and Tobago specimens and also occurs in Puerto Rico on Isla Vieques (Gutiérrez 2004).

Zebra roaches' bright coloration and bold behavior coincide with defensive abilities. They produce a defensive secretion, like their commonly kept relative *Eurycotis floridana*, with a primary active ingredient of gluconic acid. However, their defense seems a little less potent and they are a great deal less likely to use it in captivity. In nature the defensive secretion has been observed to be a strong deterrent to natural predators (beetles, lizards, toads, and bats) (Brossut 1983). Like *E. floridana,* the adults are easily differentiated from older nymphs by the presence of short, square front wings (there are no hind wings). One or both wings are commonly shed if the animal is handled roughly, similar to a lizard shedding its tail, though the wings do not grow back.

Females produce large, bean-shaped oothecae that are abandoned soon after hardening. The cases are usually dropped on the cage floor, but only if there are not adequate pieces of wood provided for the female to glue cases to. Egg carton is effectively useless. Gestation requires a few months. The 1st instars are able to cross over a vertical surface of petroleum jelly unless the barrier is wide and well maintained, a trait shared with most or all *Eurycotis*. Hatchlings are large and easily contained by standard mesh window screen. After the first molt, glass climbing is eliminated with petroleum jelly. Teflon paint is not a useful barrier for this species. Nymphs hatch out black—the white-and-black banding (as the common name 'zebra' suggests) begins to develop after the first molt.

Growth to adulthood takes four to six months and adults can live up to a year. Dog food is useful as a staple diet at all stages of growth. Fruits and vegetables are also eaten. Cultures fare best between 74-78° F (23-26° C). Temperatures below 70° F (21° C) shorten adult lifespan and prevent oothecae from hatching. Higher temperatures can increase growth, but can create desiccation problems (McMonigle 2007a).

This is one of the only species I have kept that visibly became easier with time. Usually the keeper must adapt to a species' needs or the stock dies out. I barely kept my stock from dying out year after year, then after about ten years they suddenly became easy to keep.

Eurycotis decipiens

Eurycotis floridana (Walker, 1868)
Florida Skunk Cockroach, ACS #95
Adult length: 24-40 mm
Glass climbing: Yes
Reproduction: Egg-laying

The Florida skunk roach lives up to its name with a very pungent defensive smell. The chemicals emitted when roaches are threatened resemble amaretto more than a skunk smell,

Eurycotis floridana, normal adult

Eurycotis decipiens in Trinidad © Martin Gamache

Eurycotis floridana, banded adults from Dr. Miller`s laboratory, 1998

but are likewise unpleasant and strong. Some say it smells like rotting palmetto. (This and *P. americana* are the two most likely species to be called a 'palmetto bug' in Florida.) This roach is found throughout Florida and adjacent areas of Georgia. This is the largest egg-laying species, with a long captive history.

Hatchling nymphs are large (>4 mm), but despite their weight can still scale over silicone with little difficulty. Oothecae hatch after two to three months, then energetic nymphs grow to adulthood in another five months. Adults seldom live longer than nine months. Multiple stock sources have emerged over the years since it is easy to find in Florida. The normal colored stock pictured here was collected in Kissimmee, Florida, in the mid-1990s (McMonigle 2010b). A beautiful stock was maintained by Dr. Miller in the biology department of Baldwin Wallace University from 1985-1999 that resembled some of the more colorful Cuban species. It was collected from a fishing dock in the Florida Keys during an annual marine life collection trip taken by the students and faculty. About half the adults of this variety had wide yellow lateral margins like *Eurycotis opaca*, but half looked like normal *E. floridana* (McMonigle and Willis 2000).

Eurycotis lixa Rehn, 1930
Hustler Cockroach, ACS #28
Adult length: 42-48 mm
Glass climbing: Yes
Reproduction: Egg-laying

The hustler cockroach is a deep black species with a beautiful glossy finish like patent leather. 'Patent leather cockroach' would be a fantastic common name if 'hustler' were not already established. No other species is such a beautiful deep and glossy black. The common name comes from the specific name, which refers to someone who follows an army to sell them things. It is easily differentiated from *E. floridana* by color, but more so by the tiny tegmina that skirt the edge of the meso and metanotum. Adults especially can excrete a surprising volume of milky white defensive fluid that smells like fluid from other *Eurycotis*, but is much less pungent. They produce a different type of milky secretion at the end of the abdomen and use their legs to smear it all over their body. Dirt and dust then stick to them like fly paper, but they clean it off to maintain the sticky outer coating.

This animal is similar in care to the other large *Eurycotis species*, but it is a little larger and a little different in behavior. It is more likely to cannibalize eggs, because they are not usually well hidden or covered (though *E. decipiens* also does not cover eggs well and usually does not eat them). Also, large individuals are more likely to be aggressive towards smaller ones than is seen in other commonly kept *Eurycotis*. This cockroach hails from Jamaica and the Florida Keys. A small group of specimens were found in Key West, Florida, in the 1950s (Gurney 1959a) with no further reports until 2014. Current stock comes from a handful of specimens found on Marathon Key (Jeon 2015).

Eurycotis lixa

Eurycotis lixa

Eurycotis opaca "Giant Yellow-Margined" with ootheca

Eurycotis opaca var. 'Giant Yellow Margined' Brunner, 1865
Giant Yellow Margined Roach, ACS #36
Adult length: 36-46 mm
Glass climbing: Yes
Reproduction: Egg-laying
This variety of *Eurycotis opaca* differs from the similar looking margined form of *E. floridana* from the Florida Keys because all the adults have margins (half the adults of the margined *E. floridana* stock had no margins) and adults have pale markings on the head and legs. It differs from the 'Sunset' variety (next) only in coloration. There are no reported cases of either throwing the wrong form. However, color is the only difference and they should have no trouble mating in captivity, so it is paramount the two varieties are not mixed.

Though not a given for varieties, this form is the same size as *Eurycotis opaca* var. 'Sunset'. The big *Eurycotis* are the largest egg-laying cockroaches consistently kept in captivity, since they can grow to nearly two inches in length. There are certainly larger egg-laying species in the world such as *Megaloblatta* spp. Sure, some *Archiblatta* spp. are supposed to be a little bigger, but they are hardly in culture. The only species ever common in the hobby that could grow a little larger was lost many years ago (that was *Deropeltis eurythrocephala*—the replacement representatives of that genus, *Deropeltis paulinoi* and *Deropeltis* sp. 'Jinka,' are much tinier than their predecessor).

Each ootheca is huge though appropriately sized for the bulky mother. Each usually hatches out fewer than two dozen nymphs after about three months. It can be very difficult to locate the oothecae even in a tiny habitat. Each ootheca is carefully pasted with chewed up dirt hardened by saliva then buried by the doting mother (McMonigle 2012e). The protective, hardened outer walls of the ootheca are similar to other species, but the eggs probably will not hatch if the mother has no substrate or improper substrate for use in constructing the additional layer of protection. Large particles like gravel or wood shavings as substrate would force her to abandon the ootheca without properly protecting it. This species (both varieties) is the only egg layer with this behavior. Other captive egg layers paste chewed up wood or paper over the oothecae and paste them to large objects so oothecae are very easy to find.

They climb glass readily but can be contained by a thin band of petroleum jelly after the first molt. Members of this genus have unusual 1st instar nymphs (hatchlings) that are not perturbed by climbing barriers. It is not a function of size, since 2nd instar *E. decipiens* are smaller than *E. opaca* 1st instars but do not retain the ability. A tight fitting or screwtop lid with screen glued in is suggested.

Nymphs start out a dark, blackish maroon, developing a reddish color and yellow margins after they have molted three to five times. They will eat dry dog food and most fruit. It takes about five months for them to reach adulthood and the cycle starts over again. After the last molt adults tend to live less than six months.

This is a beautiful variety that even breeds well for beginning blatticulturists. It is large and pretty, but the striking colors advertise the odiferous chemical protection with which the species is outfitted.

Eurycotis opaca var. 'Sunset'
Sunset Cockroach, ACS #96
Adult length: 36-45 mm
Glass climbing: Yes
Reproduction: Egg-laying
The adult of this variety of *Eurycotis opaca* may be the most spectacular member of the genus thus far kept in captivity. It has a white

Eurycotis opaca "Giant Yellow-Margined"

Eurycotis opaca "Giant Yellow-Margined"

Eurycotis opaca var. "Sunset" adults

Eurycotis floridana and *E. opaca* nymphs are hard to tell apart.
E. opaca "Giant Yellow-Margined" nymph

head and light yellow markings set off against a dark burgundy background. This species is large, as big as *Eurycotis floridana*. In addition to impressive size it rivals the prettiest member of the genus, *Eurycotis decipiens,* with a different beautiful dark and light color pattern.

According to various lists this variety was originally collected in Escaleras de Jaruco, Cuba. This species is endemic and found throughout Cuba, while adventive reports for Florida, USA, are likely improperly identified *E. floridana* (McMonigle 2012b). The common color variety of this species (the 'Giant Yellow-Margined') is handsome, but does not display yellow in the center of the pronotum. It has yellow margins on each side which is commonly seen on nymphs of our large *E. floridana* and has also been documented in the adults of at least one geographic variety found in the Florida Keys (McMonigle and Willis 2000). However, the Keys *E. floridana* margined variety does not have a white head—only about 50% of adults in the captive population had wide yellow margins (the other half just looked like ordinary *E. floridana*).

Like most *Eurycotis*, the 'Sunset' adults can be differentiated from late instar nymphs by the presence of a short pair of rectangular forewings (there are no hind wings). The small wings do not seem to have a purpose, although they serve an autotomous purpose in *E. decipiens* as they can be shed if the animal is handled roughly. The wings are approximately the same size for males and females, which look similar overall. Determining gender requires a look at the ventral surface of the abdomen to check for the presence in females (or lack in males) of valves used in ootheca creation.

Eggcases are large and are probably impossible to differentiate from *E. floridana* in structure. However, females are much more deliberate in their laying duties and nearly always paste substrate to the ootheca surface (*E. floridana* rarely do this in the same captive habitats). If there is an inch of substrate, the egg cases will be buried and pasted into the substrate causing the new keeper to grow frustrated with a perceived lack of egg production. Oothecae should be kept somewhat damp at all times and will take three to four months to hatch at about 73° F (23° C). Hatchlings are dark blackish with a tinge of maroon (identical to *E. floridana* hatchlings in size and color). The side margins develop after a few molts, while full coloration is generally achieved following the molt to adult.

Temperature and feedings requirements are not specialized. Nymphs grow and do well from 65-86° F (18-30° C). Oothecae probably will never hatch if the temperature stays below 70° F (21° C) for too many days.

Those familiar with other members of this genus will be aware that all stages climb glass and first instar nymphs are capable of walking over a petroleum jelly barrier as it offers little impediment to climbing. Still, the barrier works after the first molt. This species desiccates readily, so hatchling escapees die rapidly in the dry atmosphere of the normal home or structure. Escape, however, is easily thwarted by use of a bucket cage or other suitable habitat that offers screened ventilation and no gaps for the large hatchlings to squeeze through.

The *Eurycotis opaca* var. 'Sunset' cockroach is an impressive animal, easily cared for, and a fine addition to any hobbyist's collection.

Eurycotis taurus Rehn and Hebard, 1927
Minitaur Roach, ACS #37
Adult length: 18-22 mm
Glass climbing: Yes
Reproduction: Egg-laying
Eurycotis taurus are small and dark brown to black in color. Unlike the other *Eurycotis*, the

Eurycotis taurus adults

adult's tegmina are missing rather than reduced. Adults live less than a year. This species hails from Cuba. Stock was widely available in 1998, but it has not been seen since 2004.

Gromphadorhina grandidieri Kirby, 1904
Tiger Hissing Cockroach, ACS #38
Adult length: 40-75 mm
Glass climbing: Yes
Reproduction: Live-bearing
The tiger hisser is the most extraordinary of the large cockroaches in terms of fantastic coloration. It is similar to the common hisser in habits and body design, but is far more striking in appearance. The warning coloration makes it a stunning display animal, especially as adults often sit out on cage decorations. Each abdominal segment is banded in black and white (males) or black and yellow (females). The meso- and metathorax are black with a burgundy rectangle on each side, rarely solid black. Adult coloration is not as variable as in other species, but solid black individuals show up rarely and should be removed. Though solid black specimens maintain the distinct pronotum form, cultures I have seen sold as black *G. grandidieri* have been black *G. portentosa*.

Like male rhinoceros beetles and stag beetles, very large male cockroaches can have distinct minor and major forms. Minor males often have a rounded pronotum and horns that look like little bumps. Majors have more masculine male features. The pronotum is often a different shape, usually longer, with more pronounced extensions or notches. For those with horns, the major male's horns are wider or longer relative to the size of the pronotum. They often have more points or bumps at the end. Of course even within species hosting distinct minor and major formations there is variation.

Gromphadorhina grandidieri

This species averages smaller than large stocks of *G. oblongonota* and *G. portentosa*, but very large males are seen from time to time. Unlike most *Gromphadorhina*, it is not possible to estimate the size of a male *G. grandidieri* from the shape of the horns (in a photograph). Adult males have similar horn structure to *G. portentosa*, but without a major horn formation. Females are nearly indistinguishable, but the tergites are slightly more

Gromphadorhina grandidieri

granular and the surfaces inbetween more glossy. The segment above and in front of the jaws has very few small setae vs. *G. portentosa*. The front of the male's pronotum has a shallow notch so the fore edge of the head may be visible from above. However, the notch is never very deep like those on major male *Aeluropoda*, *Gromphadorhina* sp. 'Princisia big', or *Leozehntnera*.

Temperatures of 74-86° F (23-30° C) are best for reproduction. Apple slices should supplement kibbled dog food or equivalent. Basically, the care and habits are identical to other members of the genus—but the devil is in the details. Other hissers can usually breed well without fruit as part of the diet, but not these. Low temperatures will result in zero offspring. Adults are a little longer lived than other *Gromphadorhina*, often living three to four years.

Current stock came from Madagascar in 1995 and has done quite well, but is still relatively delicate. Specimens traded as 'Princisia Black and White' or 'Princisia Tricolor' in Europe may be the same species, but they showed up after 2000. It would be interesting to discover if these new stocks were independently imported from Madagascar, renamed, or even hybridized. Unfortunately stocks can be impure since it breeds readily with the common hisser, *G. portentosa* et al.

Gromphadorhina oblongonota

Gromphadorhina oblongonota Herrewege, 1973
Wide Horn Hissing Cockroach, ACS #39
Adult length: 45-80 mm
Glass climbing: Yes
Reproduction: Live-bearing
This handsome species looks very similar to the ever-present *G. portentosa*, but is relatively new to captive propagation (2007?). *Gromphadorhina oblongonota* hails from southern Madagascar. In the United Kingdom this is currently the most common species, as their *G. portentosa* stocks died out some years back. In North America and continental Europe *G. oblongonota* is much less common but available.

Nymphs are impossible to differentiate from *G. portentosa*, but they are usually darker and often have a large white patch on each side of the mesonotum and metanotum in the late instars. Similar 'shoulder' markings are lost on *G. portentosa* after the early instars and are more see-through than white. Adult females look similar but *G. oblongonota* have a darker maroon body color. The species name ('oblong notum') refers to the elongate pronotum which is the only difference between the two species and is only visible on mature males (only major males uniquely). Minor males lack an elongate pronotum shape, but have a notably upturned lip at the front of the pronotum, and the angle of the horns is slightly wider. The attractive feature of this species is the size and preponderance of major males. Majors are similar in size to the biggest of the true giant stock *G. portentosa* (~5% of giant stock and 0% of normal stock grow as large as this species is capable). Mature *G. oblongonota* males can be tiny, but 20-30% of the males in a colony can be majors. Though not a specific character, major males have elongate, pointed horns that look different from *G. portentosa*, which normally have globular, rounded horns. (Bear in mind the *G. portentosa* type specimen has pointed horns and the blunt horn structure is only 'normal' for available stocks.) As the common name suggests, the horn spacing on major males is notably wider than for other members of the genus.

Gromphadorhina oblongonota nymphs often have white shoulders in the late instars.

Females tend to be longer than big males because they have a bloated abdomen, not because any of the body parts are actually larger. As with every wingless species, the telescopic nature of the abdomen can lead to arbitrary length measurements. It is possible to overfeed smaller specimens (though not quite as easy at

Gromphadorhina oblongonota

Gromphadorhina oblongonota pronotum of large male

it sounds) to give rise to rather impressive measurements that are not 100% false. If any of the white intersegmental membrane is visible, a total length measurement is underhanded and should not be taken. Even when not measuring moribund animals, it can be easy to lose or gain 20% by the way the hisser holds itself. Most false measurements were not made through trickery—simply, no measuring device was employed. Add approximately 25-50 mm (1-2″) to all measurements provided here for visual estimates.

Care is similar to other hisser species, as they prefer good ventilation and fare well on dog food and fruits such as apple. However, this stock is slightly less accepting of lapses in care. Females commonly have two to four dozen young. Adults live two to three years after reaching maturity.

This species should never be kept with other members of the genus *Gromphadorhina* or any specimens labeled 'Princisia' (McMonigle 2008d), since the chance of hybridization is 100%. Unfortunately, much of the stock currently available has accidentally been crossed with *G. portentosa* or 'Princisia.' If adults range outside the dark burgundy color (especially if they are all very dark) or the pronotum dimensions vary unrelated to size, the stock is almost certainly mixed.

Gromphadorhina portentosa (Schaum, 1853)
Madagascan Hissing Cockroach, ACS #40
Adult length: 40-79 mm
Glass climbing: Yes
Reproduction: Live-bearing

The Madagascan hissing cockroach (hisser) *Gromphadorhina portentosa* is easily the most famous pet insect and the king of the pet cockroaches. It has long been the most popular and has been kept in nearly every country. The list of movies this critter has appeared in is longer than the list of most movie stars. Everyone has heard of the hit show where people eat or are buried in hissers on their quest to win prizes, or of hisser-eating contests. It could be argued that hissers have helped promote small pet keeping more than any other invertebrate, because many insect hobbyists began their journey with this animal. On the flip side when these 'new' pets were featured on a television news station in Florida, a number of people armed with 'cockroach hysteria' and hateful ignorance called their local officials in fear. The political backlash resulted in the Florida arthropod permit requirements in the 1990s. Hobbyists in the United States have been keeping *G. portentosa* since at least the mid-1950s and to this day there are no confirmed feral populations. (Reports have turned out to be either the native *Cryptocercus punctulatus*, which looks very much like a small hisser, or *Eurycotis floridana*, another wingless native, which does not look much like a hisser.) Even so, adventive *Gromphadorhina* would do no harm since they are neither structural nor agricultural pests. This species came to Europe (UK) from U.S. captive stock (Beccaloni 1987) which in turn came from Europe in the 1950s (originally from Madagascar of course).

All described hisser species are found on the largest island in the world, Madagascar. *Gromphadorhina* means 'snout horn,' or more literally, *gromphado*= 'like a sow's', *rhina*= 'horn.' Original descriptions were based almost entirely on the shape and relation of the males' horns and the pronotum shape. Described differences in facial setae and tergite granulation when mentioned were not quantified (no specified range of numbers) and these often vary within the same stock. Unfortunately all the *Gromphadorhina* have matching genitalia and can interbreed. For decades many of the culture stocks of *G. portentosa* were labeled

Gromphadorhina portentosa male with massive, rounded horns, giant stock (2007)

Gromphadorhina portentosa adult males from the same colony at very different sizes

G. picea (McMonigle 2003). What I find especially strange is the occasional cultures labeled as *G. picea* were not even the black strain of *G. portentosa* (*picea* means 'black,' or more literally 'pitch black').

Portentosa refers to the hiss, which is a portent or warning. These animals employ a combination of defenses to ward off predators. A loud hiss is created by air rushing through the spiracles as the abdomen is contracted. The hiss is a warning of unpalatability. If molested further, the roach will spit out large drops of fluid and jab the attacker using the strong spiny legs. The spines can pierce human skin, giving rise to the bite myth. At worst, spines may draw a barely visible speck of blood. If the attacker is not stopped by these defenses, the thick armor presents a further challenge.

Hissers are easy pets and (despite defenses) are easy to handle. They require minimal space, food, and time, and do not carry diseases associated with vertebrate pets. They make great pets for kids because they survive with little attention—and animal rights groups have not come after parents if the child neglects or kills this type of pet. They are not easily smashed and legs do not fall off easily, as seen in pest roaches.

Every container imaginable has been used to house hissers. Large nymphs and adults are easy to contain, while babies are another story. Sealed containers work well but they must have good ventilation. Hissers do well in dry or moist habitats, but die from high moisture and inadequate ventilation. Though unnecessary, nearly any substrate is acceptable. Horizontal hiding areas like cork bark and egg carton should be provided or the roaches may spend all their time in the corners and creases of the lid.

The staple diet is dry kibbled pet foods including dog and cat food. Nymphs fed high protein ferret food will be darker in color at adulthood. Certain color-enhancing fish food pellets can result in dark maroon coloration. Smaller adults with yellow abdomens can mature when fed only cat food, so protein may not be the key factor. Unfortunately there are no studies on which brands of which food cause what color, and most do not make a difference. Once maturity is reached, they do not change colors no matter what they are fed. Hissers enjoy most fruits and vegetables as well as lettuce, rose, blackberry, and other leaves. This species can survive on certain types of leaves alone, but a lot of frass is produced since leaves are primarily inedible.

Young reach adulthood in four to seven months, but can take longer. Major males molt the same number of times as minor males, but can take a few more months to mature. Adults live from two to five years, but rarely more than three. It is possible for an individual to survive six and a half years from birth to death, but three is a more common total lifespan.

Some *G. portentosa* stocks give birth even as specimens in the colony are dying from poor conditions (including one of the black stocks). On the other hand, other hisser species and some stocks are difficult or impossible to breed without good conditions. Gestation of the ootheca only takes two months, but females can abort it or hold onto it for years. If a culture has not produced babies in a year and the hobbyist improves conditions, every female might give birth within a few weeks. Food and water should always be available. Higher room temperature can be the necessary factor. A warm light source can be more helpful than a general increase in temperature. Certain foods can convince females to release their young, especially rose leaves, apples, peaches, and melon rinds. Obviously a last resort, females spontaneously abort when being eaten headfirst by a predator. If the eggs are developed, she gives

Gromphadorhina portentosa measurements can be deceiving. These adult males are nearly the same length but are not remotely the same size.

birth while being eaten. This last one is probably not a real answer to inducing captive birth.

A temperature range from 60-95° F (16-35° C) can be survived, but 74-86° F (23-30° C) is best. Admittedly, some *G. portentosa* stocks can grow and produce offspring successfully at 65° F (18° C). A basking light can be placed over one side of the cage to provide increased heat. This lets the animals regulate their own temperature and moisture. Bulbs should not be placed directly over the entire cage.

Adult males of most cultures are around 45-55 mm, but males of some cultures measure in at 65-79 mm. As with all animals, both genetics and nurture greatly affect adult size. Overcrowding and stagnant or overly dry conditions produce mostly tiny adults. (If they come running when watered or fed, they may be overcrowded or neglected.) Tiny adults mature the quickest. Feeding either dog food only or a variety of foods has only a minor affect on adult size. Though not comparable to large 'giant' stock males, normal stocks can produce rather nice males under good conditions. Even under optimal conditions, some males will be barely an inch-and-a-half. The tiny size suggests they are still young, but the ridge between the horns signifies adulthood. They will not molt again and should be removed from the culture if size is important to the keeper.

Gromphadorhina portentosa has been in culture since the 1950s and distinct culture stocks exist. Usually stocks have been selected for fast, easy breeding or color and size. For

Gromphadorhina variety commonly labeled "Princisia big"

obvious reasons, fast breeding stocks are small and the most common. One stock exists where each tibia is bent backwards; it is a sorry stock that could not have been selected for on purpose. Nymphs of any stock can be black. However, only certain stocks produce black adults other than as exceptional rarities. Males of common stocks range from 35-65 mm, females a little longer. (Keep in mind, length measurement of hissers is extremely subjective because the abdomen expands and contracts and can become unnaturally distended in life and death.) Giant stock adults grow longer and are much more heavily built. The most visible differences in giant stock include a wide body, massive head, thick legs, and a broad pronotum with horns that look more like horns than bumps. Major males have horns that are not only larger, but show allometric growth progression. For example, a large male near maximum potential size can be 10% larger, but have 20% bigger horns. The major horn formation includes massive globular horns, which can be seen in a photograph lacking a size reference.

Black stock and 'giant stock' produce a percentage of reversions, midgets, and deformities that should be removed to perfect or strengthen the strains. Reversions in black stock are very rare and easy to single out. Small males of the 'giant stock' can be picked out by the development of a ridge on the pronotum, though it is difficult to tell between an immature or small

Gromphadorhina variety commonly labeled "Princisia big"

female. The offspring of major males are most often minor males, but if the major males are consistently removed only the tendency for minor males remains. Size is influenced by hierarchy and contact with other specimens, so isolated giant stock specimens grow notably larger. Isolated specimens tend to grow more slowly and take longer time between molts.

'*Princisia vanwaerebeki*' *Gromphadorhina* sp. This complex of *Gromphadorhina* geographic races or subspecies has male genitalia matching *G. portentosa* and is only distinguishable from that species when major males are compared (McMonigle 2003, 2011e). Minor males and females are virtually indistinguishable, while differences in the adult setae and punctuation are variable and difficult to assign. Horns of the major males seemed to be placed further back on pronotum due to differences in allometric development and taper to a point in 'Princisia' whereas they are bulbous in major *G. portentosa*. However, the type specimen of *G. portentosa* has pointed horns. 'Princisia Big' and 'Princisia Black' probably first showed up in Germany around 2001. 'Big' is a uniquely colored animal with a strongly notched pronotum, whereas the other looks very much like a standard *G. portentosa* except the front of the pronotum is slightly notched. 'Princisia

Black' is nearly indistinguishable from normal *G. portentosa* but the horn structure is a little different. Other stocks called 'Black and White' or 'Tricolor' seem to be indistinguishable from tiger hissers and might be U.S. stock that came back from Europe with a fancy new name.

It may never be possible to determine exactly what these varieties were since many stocks have been tainted by *G. portentosa* or *G. oblongonota* hybridization. It is more difficult than people realize (even those who are concerned enough to know or care) to eliminate crossover contamination while keeping these species in cages in the same room. The gene or genes which cause the notch in the pronotum might be recessive, since this feature becomes severely lost or reduced in mixed cultures after the first generation and resurfaces after another generation or two. As a further complication of assigning the notch, old *G. portentosa* stock (even before the new *Gromphadorhina* were maintained in captivity) threw notched individuals on very rare occasions, possibly a result of the same recessive character.

Gyna bisannulata Hanitsch, 1950
Patchwork Cockroach, ACS #42
Adult length: 17-28 mm
Glass climbing: Adults
Reproduction: Live-bearing
Patchwork roaches are slow growing and can be very difficult to breed in captivity. The nymphs are incredibly energetic burrowers that should be given at least one half inch of substrate, such as coco fiber or peat, to dig through. Adults are as energetic at nymphs, but they can climb nearly anything and are skillful and powerful fliers. Although this species does not grow as large as *Gyna capucina*, the yellow-spotted, brown nymphs are virtually identical at the same size. Adults are easy to differentiate by color alone since the front of the pronotum has a white margin and the wings are a patchy, purple color.

Gyna bisannulata nymph

Like many members of the genus *Gyna*, female *G. bisannulata* have the ability to produce hundreds of tiny young in a single litter (though usually less than a hundred). Large numbers have a trade-off. The nymphs are tiny compared to the adult female. Getting the female *G. bisannulata* to produce a single offspring is often difficult. Small slices of fresh peach can help convince pregnant females to drop young. When successful, the survival rate of first instar nymphs can be low, but after a few molts they are very hardy. Females take nine months or so to reach adulthood (males can mature in as little as six months). Adult females live another five to six months, while males only live about three. If only a single small group of siblings is available as founding stock, it is very possible the males will mature and die before the females are ready to mate.

Patchwork cockroaches hail from Cameroon and Congo, Africa. Current stock is not new; it was traded in the 1990s under the junior synonym '*Gyna henrardi*'.

Gyna bisannulata

Gyna bisannulata female forming ootheca

Gyna caffrorum

Gyna caffrorum nymphs

Gyna caffrorum (Stal, 1856)
Chrome Cockroach, ACS #113
Adult length: 24-38 mm
Glass climbing: Adults
Reproduction: Live-bearing
The chrome cockroach is among the most easily reared members of its genus. Survival rate of the tiny first instar nymphs seems high, but there are always just so many that it is hard to tell. Once a colony is established, the substrate teems with movement even when food or water is not being added. Like *Gyna lurida*, large batches of nymphs are common, but they make poor feeders for any predator that needs substrate. Also, *Gyna* should never be fed to centipedes since they hide indefinitely and take bites out of the centipede when it molts.

The adults look similar to the standard form of *G. lurida,* but there are usually two noticeable black spots near the center of the tegmina. While the *Gyna* male always averages much smaller than the female, the male of this species is especially tiny relative to the female. Also, the nymphs of both species look similar with black-and-white markings on the pronotum, but the markings on *G. caffrorum* have expanded black areas towards the front on each side. This species is found in the southern portion of Africa: Angola, Botswana, Mozambique, Namibia, and South Africa. It showed up in captive culture as late as 2011.

Gyna capucina adults and nymphs

Gyna capucina Gerstaecker, 1883
Pink Cockroach, ACS #41
Adult length: 28-44 mm
Glass climbing: Adults
Reproduction: Live-bearing
This is the largest *Gyna* species maintained in captivity. The shiny, torpedo-shaped nymphs burrow and are rarely seen. Nymphs are chocolate brown with twenty or so large yellowish-tan spots. The diamond-shaped pronotum of the adult extends partly over the abdomen and is painted a dark maroon with some lighter markings. The outer wings (tegmina) appear pink, but upon very close inspection it is possible to see extremely thin alternating bands of red and white. The antennae are banded in orange and black. The underside of the adult is a peculiar orange-pink color.

Immatures grow to become adults in six to nine months at 75° F (24° C). I tried keeping this species at 80° F (27° C) to improve production—it killed off the nineteen-year-old colony over about a year, though there may have been other factors at play. Although some losses are commonly seen in the first instar, especially if crowded, the only truly difficult part of keeping this species is getting females to produce young. As with *G. bisannulata*, the inclusion of a few small, fresh peach slices seems to help. They eat dog food and seem to develop better if some rotten leaves are included in the substrate. Females are mostly much larger than males, but some males can be pretty large so genders can overlap in total length.

Gyna capucina

Gyna capucina was the first cockroach article in the premiere issue of *Invertebrates-Magazine* (McMonigle 2001). Current stock came from specimens kept at university laboratories prior to 1995.

Gyna centurio Dohrn, 1888
Centurion Cockroach, ACS #130
Adult length: 19-28 mm
Glass climbing: Adults
Reproduction: Live-bearing
This handsome cockroach is similar in breeding to *G. lurida* and has proven rather productive. The nymphs are very easy to tell from other commonly maintained species since they are dark brown with around half a dozen yellow spots towards the front. This species hails from Cameroon and Congo, Africa. *Gyna centurio* showed up in culture before the turn of the century (Durkin 1998b).

Gyna species are members of the family Blaberidae, live-bearing roaches. Oddly, their minimal appetite, food preference, adult shape, burrowing immatures, and habits resemble the polyphagids (desert cockroaches). Females retain eggs in the body until they are ready to hatch. Newborns are extremely small in relation to the adult, which allows them to produce larger clutches than any other cockroach genus, up to 200+ nymphs at one time.

Gyna centurio nymph © Pascal Garbe

Gyna centurio nymph © Pascal Garbe

Gyna centurio adult pair and some banana © Alan Jeon

Gyna lurida Saussure, 1899
Porcelain Cockroach, ACS #43
Adult length: 20-35 mm
Glass climbing: Adults
Reproduction: Live-bearing
This elegant cockroach comes from Kenya and Tanzania, Africa. The energetic nymphs are mottled in black and white and spend their time burrowing through the soil. They do not climb smooth surfaces. Porcelain cockroaches are a pretty species with calico nymphs and variable adult coloration—often a mottled, light gray color like porcelain. Adults are active during the day. Adults can burrow like nymphs, but often stay above the surface. The overactive adults climb glass and are one of the only roaches capable of flying straight up from the bottom of the cage. Males are generally small, a little over 20 mm, and usually an unattractive blotchy brown, but occasionally yellow. Females are far more variable in size and color. They range from 25-35 mm. Most adult females are a delicate porcelain color with variable brown markings, while some have bluish spots and others are orange or an incredible lemon yellow (McMonigle 2004a). I forgot there was an orange form, because the yellow is so spectacular (the blue markings have more to do with proximity to the molt). It can be difficult to get the yellow to come out as bright in

Gyna lurida standard with nymphs (2000)

Gyna lurida, yellow form female (2002)

photos as it is in real life. The beautiful yellow form was originally very rarely seen in cultures, maybe just once. Then again, the word *lurida* means yellow. I spent multiple generations and years isolating from the first yellow female so we now have a stock that breeds true. Only the previously porcelain-colored areas of this variety are yellow. The normal patches of brown to black pigment remain and can vary greatly by individual.

The enclosure should be screened at the top for ventilation. Newborns will squeeze through normal window screen but do not escape as long as the screening is not located near the substrate. Most substrates are acceptable, though a few inches of decaying leaves (at least on top) are preferred. A water dish (filled with sponge to prevent drowning) should be in the cage. The tops of the dishes should be flush with the substrate because young nymphs are small and cannot climb. While water is important, porcelain roaches can live through very dry conditions. In addition to the dead leaves, food includes processed pet foods and fruits. Even large cultures do not eat much, so a small food dish is used. *Gyna lurida* are prolific and few hobbyists have trouble breeding them. In contrast, the other more established *Gyna* in the hobby (*G. capucina* and *G. bisannulata*) can be rather difficult to get babies from. This is the newest of the old cultures, which showed up, at the very latest, in 1997 (L.M. Roth identified specimens).

There are other day-active cockroach genera like *Pseudomops* and *Eurycotis*, but they are active mostly dusk to dawn. *Gyna* species are the only arboreal day-active roaches available in the hobby. Nearly every other pet cockroach escapee will be found dead and dried somewhere under furniture or in a crack days or months later. The flighty adult *G. lurida* flitter away now and then, but they will be found sitting on a houseplant, shelf, or wall the next day. They are swift, but not too difficult to catch.

Hemiblabera tenebricosa Rehn and Hebard, 1927
Broad Keys Cockroach, ACS #90
Adult length: 36-42 mm
Glass climbing: No
Reproduction: Live-bearing
These resemble the Cuban *Byrsotria* species in most ways, except the males have very short wings like the females and they are more reddish-brown overall. The name broad Keys cockroach dates to Helfer's 1954 guide or earlier, and refers to the wide body and the presence of this species in the Florida Keys. Sometimes this species will be called the horseshoe crab or trilobite cockroach like most of the *Byrsotria* and *Aspiduchus*. The protected *Aspiduchus cavernicola* from Puerto Rico looks nearly identical to this species, but the wings are a little longer on the female.

Hemiblabera tenebricosa has been used as a feeder and reproduces very easily. Nymphs and adults do not mind a lot of tank mates and it is very difficult to get to a point where problems arise from overcrowding. Females refuse to give birth to young at 72° F (22° C) or lower. The eggs probably cannot develop at the lower temps, though in cold conditions no abortions have been noted. Nymphs still feed and grow at even lower temperatures.

Hemithyrsocera sp.
Asian Clown Cockroach, ACS #128
Adult length: 9-12 mm
Glass climbing: Yes
Reproduction: Egg-laying
These spectacular little critters from Southeast Asia have been maintained in Europe since at least 2012 and stock is currently labeled *H. histrio* (Burmeister, 1838). They began to show

Hemiblabera tenebricosa

Hemithyrsocera histrio adult © Pascal Garbe

Hemithyrsocera lateralis. Malaysia (CC-BY) Tomas Maul

Hemithyrsocera histrio
subadult © Pascal Garbe

up in U.S. collections in 2015. By itself, a nymph or adult is a handsome little animal (though nymphs and adults look entirely different), but there are many other colorful and quite small species of cockroach. However, a large group of clown roaches makes an impressive display. In a crowded enclosure they run out of hiding spots and sit out in the open. This is a highly productive species and it will be interesting to see how it fares with the inevitable crash in demand. Cyclical value based on supply and demand of extremely productive and easy to produce species is why some species are lost to culture. If one or two keepers maintain stock through what could be a very long period of zero value, the cycle can start over again, if the keeper trades or sells them.

Henschoutedenia flexivitta (Walker, 1868)
Giant Lobster Roach, ACS #44
Adult length: 30-38 mm
Glass climbing: Yes
Reproduction: Live-bearing
The giant lobster roach may be more closely related to *Rhyparobia*, but the adult coloration looks a lot like *Nauphoeta cinerea* (hence the common name). The handsome nymphs look completely different and are striped with light

Hemithyrsocera histrio adult
and hatchlings © Pascal Garbe

margins where the segments join. Contrasting stripes usually serve warning to predators, but the nymphal defense is a seemingly minor odor to human senses. Possibly certain birds do not like the flavor.

Henschoutedenia flexivitta is a large soft-bodied species with limited defenses and can be a useful feeder for lizards and invertebrates. Specimens are easily reared and produce in quantity if offered fruit regularly, but colony crashes are likely if given limited attention. Adults and large nymphs are prone to dropping over dead from neglect. Cannibalism is another side effect of neglect. It is not difficult to maintain the stock for decades, but it takes consistent care to maintain a large feeder culture.

Colonies of this species have never been readily available, though it has been on vendor lists back to 2000. It was one of the few species on the 1986 Blattodea Culture Group

Henschoutedenia fexivitta

Henschoutedenia fexivitta

Henschoutedenia fexivitta

stock list. Wild specimens are found in western Africa, including Cameroon and Ghana.

Ischnoptera deropeltiformis (Brunner 1865)
Dark Woods Cockroach, ACS #48
Adult length: 12-19 mm
Glass climbing: No
Reproduction: Egg-laying
The dark woods cockroach is shiny black with reddish-brown legs. Adult males fly, while the females cannot since their wings stretch only about two-fifths the length of the body. The genus name means 'thin-winged' and may reference the venation. The species name suggests this cockroach is somehow similar in form to the blattid genus *Deropeltis*. *Ischnoptera deropeltiformis* belongs to a different family (Ectobiidae), females are not wingless, it does not produce quinones, and is relatively tiny.

In captivity, *I. deropeltiformis* do not glue debris over oothecae—in nature the oothecae can be found in moist dirt under loose bark. Adult males live three to four months and females live seven to eight months.

The dark woods cockroach is common across the eastern United States in wooded habitats similar to those where *Parcoblatta* spp. frequent; it is sometimes found under the same logs. *Ischnoptera bilunata* (Bilunate Roach ACS #132) is found from Florida to Texas. Nearly a hundred species of *Ischnoptera* are found in the Americas; the genus is almost entirely tropical. Another small, light reddish-brown *Ischnoptera* species from Costa Rica (Strawberry Cockroach ACS #114) is an occasional feeder that showed in in the last couple of years.

Latiblattella lucifrons Hebard 1917
Santa Rita Cockroach, ACS #138
Adult length: 15-18 mm
Glass climbing: Yes
Reproduction: Egg-laying
This small, frail species is found only in the Santa Rita and Huachuca mountains of Arizona. Males have long wings and females have shorter tegmina. They can be found near the top of understory vegetation late in the evening. Both sexes are attracted to lights. This is one of our few natives that is almost never seen in captive culture.

Latiblattella lucifrons

Ischnoptera bilunata, female with ootheca, found under leaf litter near the base of a tree, Montgomery, AL © Alan Jeon

Ischnoptera deropeltiformis, adult female, found under leaf litter on the forest floor Montgomery, AL © Alan Jeon

Ischnoptera deropeltiformis, adult male, found under leaf litter on the forest floor Montgomery, AL © Alan Jeon

Loboptera decipiens (Germar, 1817)
Lobe-Winged Cockroach, ACS #135
Adult length: 8-10 mm
Glass climbing: Yes
Reproduction: Egg-laying
Europe's colorful, tiny, native cockroach ranges across the southwest countries, but also through Northern Africa and the Middle East. The 'lobe wings' are the tiny little reduced tegmina on the sides of the mesonotum that are barely visible without magnification. Adults are shiny, dark brown animals with bright yellow margins around the entire body. A striking feature is the size of the ootheca, because it can be nearly as big as the female that made it. Often twenty to twenty-six eggs are contained within, and the hatchlings are relatively large. Oothecae and animals should be kept damp at all times since they dry up rapidly, but they will die if kept too soggy. 1990s culture stocks were originally collected in France and Greece and looked identical. Greek specimens have been collected in old conifer forests under rocks and around the bark of dead pines (Turner 1995). This species can also be found around rocks and seaweed by the shore. It is not a terribly difficult species to keep, but it never makes it past a few generations because care requires regular attention and it is tiny. (Humans lose interest in things that are difficult to focus on.)

Loboptera decipiens adults, 1990s stock

Lucihormetica subcincta (Walker, 1868)
Glowspot Cockroach, ACS #45
Adult length: 29-40 mm
Glass climbing: Adults; nymphs less so
Reproduction: Live-bearing
The name glowspot comes from the pair of raised yellow spots, found only on the male's pronotum, that are supposed to glow in the dark on wild-caught specimens. The spots are not exoskeletal pigment, but are inflated organs located just below the surface of the exoskeleton among the horn structure. The exoskeleton covering the organs is transparent and very thin. It can collapse when the organs dry out on preserved specimens. Old pronotums left in the cage usually end up with holes where the organs were.

Lucihormetica subcincta started out as *Brachycola subcincta* and was later revised as a member of an earlier described genus, *Hormetica* (1838). In 1999, *Hormetica subcincta*, along with another eight members of the genus *Hormetica* (fifteen species presently remain in the genus *Hormetica*), were moved to a new genus, accompanied by the description of a new species, *Lucihormetica fenestrata* by Zompro and Fritzsche. Four new species have been described since. The last, *Lucihormetica luckae* Vršanský et al., 2012, was endemic to an Ecuadorian volcanic region and appears to have been unlucky. Eruptions of the local volcano may have killed every last *L. luckae* specimen.

The species of *Hormetica* moved to the genus *Lucihormetica* all have one characteristic in common: the adult male only has a pair of luminescent organs near the middle of the pronotum. The prefix '*Luci-*' means 'light.'

Loboptera decipiens: These were collected in the central coastal plain of Israel in 2005, in husbandry ever since. They prefer dry conditions, and get their water from food. They feed on fruits and vegetables, with dog food and fish food as supplements. They enjoy eating leaves as well. © Gil Wizen

Loboptera decipiens feeding on baking yeast. Captive-bred stock, the stripe is genuinely pure white. The colony is given a relative humidity of 70%. The temperature gradient is lowered to 14-22° C from December to March, and raised to 21-26° C in summer. Their favourite foods are carrots, salad leaves, fish food, and cereals. 2016 © Bill Ogden

Lucihormetica verrucosa eating dog food

Lucihormetica subcincta adult pair and nymph

These are the only luminescent orthopteroids known. Unfortunately, dead specimens and captive-reared individuals lack this ability. Still, the idea is a rare chemical or luminescent microorganism may be missing from the captive diet. Dissection and microscopy of the glow organs suggest the ability to glow under its own power may be science fiction (Grevan and Zwanzig 2013). The "glow" organs do glow orange under ultraviolet light, while other pale-colored areas of the pronotum fluoresce green. Furthermore, in captivity some diets (specifically color-enhancing fish food with canthaxanthin, less consistently pumpkin, and carrots only with some imagination) can change the color of the spots from yellow to a deep orange any time after maturity (McMonigle 2009a). A significant change in color takes a few days to a few weeks. Dietary changes often affect coloration as a cockroach exoskeleton is being formed, but a rapid dietary color change afterwards is unique. Orange spots fluoresce more brightly than yellow under UV.

Males and females are easy to tell apart as they reach maturity. The adult male is shiny and bright and the yellow organs on the pronotum have formed. The pronotum of the male is also structurally dimorphic. The female has a smooth, rounded pronotum while the male's has upraised ridges (similar to dimorphism in

These *Lucihormetica verrucosa* are feeding on apple. Adult females of this species can have multiple yellow spots in the center of the pronotum, just two, or none.

the hissing cockroaches—which were *Hormetica* before they were *Gromphadorhina*). As usual, the abdominal segmentation is dimorphic, but this is only noticeable in the later instars. The most visible sexual characteristics are not visible prior to the ultimate molt. Males are almost never observed sparring with the horns, unless one is removed from the enclosure for a few weeks and then replaced.

Lucihormetica occur primarily in the northern half of South America. *L. subcincta* is native to Columbia. It is found in tropical forests, commonly not too far from streams or other sources of fresh water. This glowspot is capable of surviving dry periods, but lack of available moisture leads to damaging molts and death. Natural temperatures normally range from 75-88° F (24-31° C) but rarely go as low as the mid-sixties. Nymphs are slow growing and can take up to two years to reach maturity—the adults live another one to four years. Although the majority do not live so long, multiple four-year-old females have been known to produce young. In other types of cockroaches, females that live overly long far outlive their reproductive viability (McMonigle 2009a).

Lucihormetica verrucosa (Brunner, 1865)
Warty Glowspot Cockroach, ACS #94
Adult length: 36-46 mm

Lucihormetica verrucosa, different sized males

Lucihormetica verrucosa climbing glass

Glass climbing: Adults; nymphs slightly.
Reproduction: Live-bearing
This species is similar to the standard glowspot but averages larger and is far more productive in captivity. Its wings are longer and rounded and variably marked in black. The dark abdomen also has a white margin as well as variable stripes between segments. A diet of about half dog food and half fruit (apple, pear, mango, cantaloupe, pumpkin, etc.) seems to work well for production. Although it seems to breed fast enough to use as a feeder, cultures take a long time to recover. Specimens are usually harmless to each other, but limited food and space can result in chewed wings and body damage. Major and minor males can differ incredibly in size and also horn structure on the pronotum.

Like other members of the genus, the male is reminiscent of the Jawas of Star Wars when viewed from above. Late instar nymphs and females of this species are variably marked. Females can have a dark central marking on the pronotum, a mosaic of yellow markings, or a small pair of spots. Sometimes it can be necessary to take a look at the last abdominal segment just to make sure an individual is not a minor male, though the spots are just pigment. Female *L. verrucosa* and other blaberids (*Blaberus boliviensis* and *Eublaberus serranus*) can have a pair of pale spots on the pronotum, but not the inflated glow organs, nor the elastic transparent windows in the exoskeleton. Since the markings are most advanced or present only on the male of *Lucihormetica*, they could play a role in the female's visual selection of a mate. How exactly the colored dots function towards reproduction as well as its relevance to this species remains a mystery, though the spots are usually larger the bigger the male is. Vršanský et al. (2012) suggest *Lucihormetica* mimic toxic luminescent click beetles (family Elateridae) also found in the tropical New World. Unlike many species with wing dimorphism, *Lucihormetica* males are flightless and live under the same logs as females. It does not seem like there would be adaptive benefit to leaving the females unprotected. If it is warning coloration, it might not mimic anything, but advertises its ability to sequester toxic compounds in the organs.

Lucihormetica grossei is a smaller, less colorful species with much longer wings. It was in culture in Europe shortly prior to its description (Fritzsche 2008), but has not yet made it to North American lists.

Lucihormetica grossei © Pascal Garbe.

Macropanesthia rhinoceros Saussure, 1895
Rhinoceros Cockroach, ACS #50
Adult length: 60-80 mm
Glass climbing: Can barely walk
Reproduction: Live-bearing
There are twenty-four different species of large tunnel-building cockroaches (Geoscapheinae) in related genera endemic to northern Australia (Rose et al. 2014). These large, wingless, live-bearing roaches are specially adapted to life spent primarily in underground burrows. (Unlike many blaberids, they build burrows;

Macropanesthia small eyes and no ocelli, unusual cerci, and burrowing front leg

they do not burrow through substrate and are uniquely unqualified for unstable subterranean life.) *Macropanesthia rhinoceros* is the largest of these species and is the heaviest recorded cockroach in the world—a gravid female can weigh over 30 grams. A small adult male does not look impressive compared to the largest major *Gromphadorhina*, but big males are truly impressive. There is a lot of variation in adult size (usually more for males than females) and 20 mm in length translates to 100% difference in bulk. The rhinoceros cockroach is the holy grail of the enthusiast, with captive-bred pairs having sold for close to a thousand dollars in recent years. This heaviest of all roaches makes up in bulk what it lacks in beauty. It is also one of the longest living roaches. Nymphs can take more than five years to reach adulthood (62 months McMonigle 2006a). Adults may live an additional three to seven years.

Rhinoceros cockroaches are found in central and north Queensland, Australia. There they live in sandy soil in stands of cypress pine (*Callitrus* sp.) and scrub areas of *Acacia harpophylla* (Day 1950). They build somewhat permanent homes in the sandy soil. Gently sloping spiral tunnels do not go more than three feet deep but can be eighteen feet in length and end in a small chamber usually filled with rotten leaves. The entrance to the burrows is sealed in the dry season. They are very common in the habitats they occur; in some areas burrows are located every square meter (Beccaloni 1990).

Macropanesthia cerci

Rhinoceros roaches have a number of curious features adapted for an underground life. They have short, rounded cerci that can be retracted into indentations in the terminal body section. This probably protects the cerci from damage during burrowing. The tarsi on all legs are highly reduced and have close to zero function. They are so reduced that when I first saw one, I thought the tarsi had been broken off from poor handling. Of course without proper-sized tarsi they cannot climb, not even over a 1″ (2.5 cm) stick (though they can go down under). The front tibiae are rounded into spiky shovels like a mole cricket's forelegs. The eyes are unusually small so the face looks more like that of the "niña de la tierra" (Jerusalem cricket) than a roach. The *M. rhinoceros* brain is a study in miniaturization. The huge head contains a tiny brain that only takes up a tenth of the space it would in another cockroach of the same size (Day 1950). The small brain may result from low processing requirements for the minimal input of reduced eyes, antennae, and cerci.

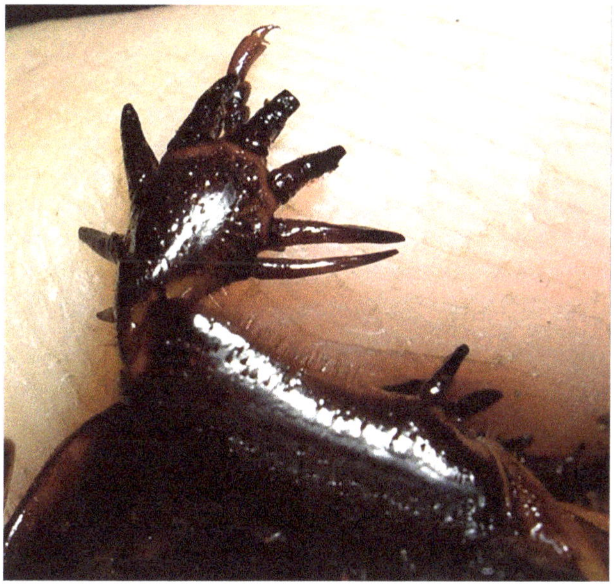

Macropanesthia tarsus

There are some semi-social interactions reported from the wild. First, babies normally remain in the mother's burrow for nine months before going off on their own. Second, adult males are usually found with subadult females, patiently waiting for the young lady's final molt; a study in patience that could be more than half a year away (Rugg and Rose 1991). In captivity both adult and subadult males have been observed fighting in underground chambers for hours on end after a female matures (McMonigle 2006a). They cannot damage each other and afterwards return to peaceful cohabitation. They often seem to choose to inhabit the same tunnel, but in captivity it is very difficult to tell social behavior from hide preference due to the confined space of terrariums. Nymphs can be kept together to maturity, but each cage should contain no more than one male and one female adult or resultant young will be smashed and eaten. Certain stages can be kept safely together but these are naturally solitary, not social animals. This can be difficult to grasp for keepers of other roaches.

Unlike termites and *Cryptocercus*, *Macropanesthia* do not die or display stress if prevented from forming shelters. Still some keepers may choose deep-substrate enclosures, especially for mated adult females. It is important to remember the substrate must either be deep and compactable enough to provide for tunnel building or it must be shallow. They are not substrate swimmers like *Byrsotria* and *Blaberus*, and nymphs rarely survive a buried molt. In deep-substrate enclosures they may never be seen by the keeper unless dug up. If maintained in deep substrate, around November every year they come to the surface and wander around the cage every night for almost a month (specimens kept in North America early 2000s). It does not matter if they are nymphs or adults, male or female. The rest of the year they stay underground most nights but occasionally can be observed searching for food.

The life cycle is very slow for a roach. The minimum time to reach adulthood is thirty-three months, while large adults take at least forty-eight months. Usually, the larger the animal the longer it takes to mature. However, poor conditions that do not kill nymphs could stretch out the time greatly, unrelated to size. Rhino roaches are reported to live ten years, but that includes the time to reach maturity. So far the longest kept animal was an adult female that lived seven years at the University of Sydney (Rugg and Rose 1991). Captive females usually have their first litter one to two years after the molt to adulthood. The brood hatches around November in Australia, near the beginning of summer. There is only one batch a year, but there can be as many as thirty nymphs.

Males and females look very different at maturity. On the last molt the male's pronotum changes shape to an upward scoop, which is where the species name comes from. He can also have a small pair of horns, but they are not as pronounced. Major males have secondary horn formation with small points towards the middle. No matter how small a male is when he develops the scoop characteristic, he is mature and will not molt again. (Small males molt the same number of times, but faster than larger males.) Both genders change color from dark reddish-brown to a lighter red-brown on maturity, though color is subjective. Females keep the rounded pronotum shape and have no horns, so it is difficult to differentiate a small adult from a large nymph unless it has had young or has not molted in two years or more. The underside of the adult female's abdomen is a lot darker and feels firmer than a nymph. Though noticeable, telling the difference comes with experience (pers. comm. Rose 2006, on wild specimens). For captive specimens, the segment hardness indicator does not seem to work, but adult color is not only lighter, but also monochromatic, without any darker segment margins.

One thing that does not seem to be common knowledge is this cockroach hisses just as do the popular hissing roaches from Madagascar. The soft hiss was documented at least as far back as Rose (1990). The noise is not as loud or readily employed, but it is very noticeable.

The Enclosure: Any plastic or glass container with a surface area of at least 10" x 10" (25 cm x 25 cm) and a height of 4" (10 cm) makes a good cage. Of course the container can be much taller but greater area can make food and moisture more difficult to find. Less than an inch (1-2 cm) of potting soil mixed with sand is good for the substrate. Deeper substrate allows them to make nifty tunnels, but the substrate will collapse on them during molts and result in massive deformities and mortal damage. Use of clay, compost, or dirt from outdoors is not desirable because these materials cake up on the roaches legs and mouthparts. Coconut huts or a piece of lightweight bark or plastic gives them a place to hide. When choosing decorations, do not forget that a 1" (2.5 cm) high barrier can block them from food and water.

Temperature: 55-79° F (13-26° C). In nature their burrows are approximately 68° F (20° C) year round (Rose 1989). High temperatures are known to kill this species.

Moisture: An area of the substrate should always be kept wet because *Macropanesthia* die without water. Small nymphs can survive a few days on dry substrate, adults far longer, but a third instar nymph died after five days without water (2014, 2nd generation). Most roaches are not so tolerant of dryness. If the substrate is overwatered and becomes soggy, the roaches wander ceaselessly in hopes of escape. In captivity, imited ventilation has not been tested since it likely would suffocate these massive insects. They are accustomed to arid habitats, but they are well protected in their burrows.

Food: In Australia, *Macropanesthia* supposedly feed on old, dried eucalyptus leaves and twigs (Rose 1990). Another report (Day 1950) states they eat grass roots and dead leaves among pine and acacia stands. In addition to fallen leaves, they gather sticks up to 15" (38 cm) long by 3/4" (2 cm) diameter. They eat the bark off these sticks (Brotherton 1986). The primary food used has been whole kibbled pieces of dry dog food offered once a week. Uneaten dog food, if present, is removed after three to four days. Old, dead, brown leaf litter from a mixed hardwood forest is offered a few times a month (as consumed). A piece of apple or melon is offered once a month though is rarely entirely consumed. Rotten wood is sometimes kept in the cage and nibbled on from time to time. One accompanying photo shows a piece of rotten wood being eagerly fed on when it was first placed in the cage.

In captivity, multiple consecutive generations (accompanying photographs) grew up on a mix of oak, elm, and mixed hardwood leaves. The weathered and dry (not green) leaves had been on the ground one to eight months. Old leaves are necessary because freshly dried leaves, aromatic Eucalyptus primarily, have been implicated in some deaths. When crushed, the leaf should break, not merely bend. Any hardwood leaves can be used, but they should have been on the ground for a season to prevent mite problems as much as to allow for the breakdown of harmful oils. Sugars and oils do not break down if leaves are collected and stored dry for multiple years.

Macropanesthia do well if fed 40-80% dog food by weight (balance leaves). If placed on a piece of dog food they immediately begin eating it, or biting the hand that recently contained dog food. Food is rarely sought out if placed out of reach in the enclosure. Dog food kibbles should be placed in a tunnel or under a hide, never in a dish. This species cannot gain access to food in a small dish. In a large dish, even if the animal could get into the dish, it would be unable to get back out and would likely dry up or end up on its back. They will die if on their backs too long (longer than 24 hours upside-down is likely to kill them). They can also eat rotten wood and nibble on a variety of fruits and vegetables.

Reproduction: In Australia, female *M. rhinoceros* are known to give birth in nature once a year, normally during the month of November. According to one reported captive birth (pers. comm. Davson) this timing can also hold true for imported adults. However reproduction tends to occur well outside this time period in captivity. A pair of adults acquired in 2000 produced young June 15th, 2002. One female (and three males) from this captive hatch were acquired the first week of July 2002. The nymphs were reared to maturity and the female produced a batch of young on July 4th, 2008. Subsequent captive births through 2015 continued to occur in June and July. November is just prior to summer in Australia while June-July is early summer in North America. The correlation suggests seasonal temperature changes trigger the reproductive cycle. (Indoor temperature shifts still reflect outdoor seasonal temperature changes, unless unique care is taken to maintain specific temperature ranges.)

Up to 30 was listed as the brood size (McMonigle 2006a). Precise data from five captive births are: 22, 23, 24, 24, and 26. (A sixth brood was almost entirely eaten by the adults and a seventh was born damaged due to excessive dryness , so they could not be accurately counted.)

Motherly attention is visible in captivity as the babies are always within an inch of the adult, often touching. This is not the protective behavior found in other species that lack sheltered burrows, as *M. rhinoceros* scatter when disturbed instead of hiding under the adult. The 1st instar young (July 2008) seemed unable to distinguish the female from the two adult males in the cage. Half the nymphs were commonly found with the males when the adults were on the surface. The female and males have cohabitated continuously since birth and would have the same odor if smell were a factor in recognition. The female formed a small burrow for the young and all of the young eventually found their way to the burrow. Within a few days the larger male located and tried to share the tiny burrow. The mother and young were removed fifteen days after birth because three of the babies disappeared—there was concern that they might have been eaten by the males. Once separated no further losses occurred for a few months. Females drive males from the burrow before young are born in the wild, but that is not possible in the close quarters of a small enclosure. Patricide was confirmed in 2014 when three males ate all but one of a female's offspring inside a small 5″ x 8″ enclosure.

The original 1st instars were removed from the mother a few weeks after birth (July 2002) and successfully matured and reproduced. The 2008 brood was removed after a few months; subsequent broods were removed as soon as encountered. Keep in mind the wild females tend to have fewer than ten nymphs in their burrows, not 24, and the male is driven from the burrow before they are born.

It can be concluded that the mother's care is not required for captive survival, but could it have an effect on overall survival or growth? Survival after removal from the mother of more than a hundred nymphs has been perfect, except for a third instar that died from desiccation. Nymphs of the subsocial cockroach, *Cryptocercus punctulatus,* are far more dependent on their parents in nature and do not grow quite as large in captivity if reared separately. It is difficult to say how the *Macropanesthia* mother's attentions affect adult size without control groups, because the food and temperature are different, but the animals kept with their mother for the first month or more were not larger than later generations. Also, a 3rd generation male matured March 15th, 2016 at 77 mm, nearly maximum size for this species in the wild.

Usually if a female matures close to the birthing season it just tacks on more time before her first brood. However, one of the 77 mm male's smaller sisters matured in December 2014 and produced young in June 2015.

Captive Maturation Time: Four of five nymphs born on the same day in June 2002 matured successfully over a prolonged period. The fifth, a male, became excessively deformed during the molt to adulthood from a collapsed burrow. Rapidly maturing specimens were very small. One undersized male (57 mm) matured in August 2005, the second (60 mm) in September 2005. The third, far larger male (71 mm) matured in March 2007. The female matured (74 mm) on August 16th, 2007. Her previous molt occurred on May 1st, 2006 (McMonigle 2008c).

The first molts are extremely rapid compared to later molts for this species. The 1st instars began to molt on July 22, 2008, just eighteen days after birth. They were offered dog food, rotten leaves, and melon rinds before this molt and consumed a minimal amount (dog food made up ~80% of consumed food by weight). While eighteen days is not rapid for most insects, it seems rapid for *Macropanesthia*—one-inch nymphs molt every three to six months, but the time between each molt grows longer with age until the final molt that can take more than fifteen months. Nymphs molted every few days and the last nymph molted to first instar on August 15th (day forty-two). A few days later the first 2nd instar molted to 3rd instar, just twenty-six days after the first molt. Later generations through 2015 underwent the initial molts at similar rapid initial intervals growing progressively delayed.

The table shows molting dates for one brood in captivity. (Time required for the first and last individual to molt are listed for 1-3, but only the first is listed for 4-8 since not all specimens were available or did not complete

Macropanesthia female with first instar nymphs underground and then dug up (7/13/2008)

Macropanesthia female with nymphs

Macropanesthia major male horns (left); nymph at right is a slow growing female sibling (2016)

Macropanesthia one-year-old nymphs (7/19/2009)

Macropanesthia subadult male and adult female can be difficult to tell apart. Normally, subadults (above male specimen) have very dark margins towards the front of each abdominal segment. Genders are approximately the same size but this male will be very large at maturity (2015).

Macropanesthia molting on surface (2010)

these molts at the time.) There were no supernumary molts. It is interesting to note that one small male matured in only 33 months, whereas 38 months was the earliest the previous generation (57 months for the largest male, 62 months for the largest female). Undersized males are always the first to mature, but late maturity does not necessarily indicate large size.

Molting Periods Table
(from McMonigle 2011b).

Born	7-4-2008
Molt 1	7-22 to 8-15-2008
Molt 2	8-17 to 9-20-2008
Molt 3	10-9 to 11-28-2008
Molt 4	1-10-2009
Molt 5	4-19-2009
Molt 6	7-12-2009
Molt 7	1-9-2010
Molt 8	6-15-2010
Molt 9	3-26-2011

Two of the 2008 brood were born with damaged rear and middle legs that were wholly to partly discarded after the first molt. Many cockroaches are not very good at growing back legs, so it was feared they would always be deformed. The legs slowly grew back and by molt five it was no longer possible to observe where

Female *Macropanesthia* with first instar nymphs.

Macropanesthia nymphs molting to second instar (7/26/2008)

Female *Macropanesthia* with second and third instar nymphs (9/13/2008)

Macropanesthia male with confused nymph (713/2008)

the original damage was. Those with damaged legs have been among the last to molt. Three from a 2014 brood had damaged legs that regenerated only after four to five molts.

Substrate has been kept shallow (around the same depth as the height of the animal) during development because deeper substrate caused massive, sometimes fatal, deformities in earlier attempts (late 1990s). Initially this was mistaken for a food issue. Although it is often called a giant burrowing cockroach, it should really be called a giant cavern-building cockroach. It constructs tunnels with a terminal cavern in nature and does not live covered in substrate like the various blaberids kept in captivity. Deep substrate is not a problem during normal activities, but directly causes massive deformities during captive molts. Shallow substrate greatly increases the risk of desiccation, but it is not that difficult to keep the substrate from drying out—and *M. rhinoceros* is less susceptible to low humidity than most roaches. The survival rate with described parameters has been 100% after removal from the adults, with the exception of one case of extreme dehydration.

Consecutive broods of pictured stock were 2002, 2008, 2012, and 2015 (with nonconsecutive sibling broods produced also in 2013, 2014 and 2016). All births took place June or July, while maturation of adults occurred in any month. It may seem like the time between generations decreased, however the first female was alone, large, and took more than five years to mature, while the last female was small and matured in less than three years (ten months before her next closest sister and sixteen months before her largest brother). This small female also produced young just seven months after maturity, which is unusually quick. Four to five years is the most likely time needed to pass one generation and most captive adults do not live more than four years. Of six captive broods from 2008 to 2016, never once did a female produce young more than a single time in her lifetime even if she lived another four years (for a total of five). Still, wild females probably produce more than one brood and the unusual 2015 female shows promising signs of a 2016 brood because she is very fat. Nevertheless, past broods have issued from thin females while their portly sisters did nothing.

While some writings suggest rhino roaches have become popular pets in Australia, there have been no real captive-breeding efforts and keeping them alive has been the challenge. "There have been too many reports of cockroaches dying" (Rose 1990). Most if not every animal besides those pictured here being kept outside Australia seems to be directly from, or one generation removed from, a single Australian insect farm (farm does not suggest indoor rearing). Captive reports of reproduction in November suggests the specimens were in the southern hemisphere (i.e. Australia) in the last twelve months. Rhinoceros roaches are probably the most desired and expensive roaches. Their tendency to eat small young unless properly spaced, slow and irregular growth, and restricted reproduction will make certain this is the only cockroach that could never be used as a feeder.

Megaloblatta longipennis (Walker, 1868)
ACS #49
Adult length: 70-97 mm
Glass climbing: No
Reproduction: Egg-laying
This has long been the most sought-after captive cockroach and yet it has proven impossible to acquire except for the hard-to-find dead specimen from Peru. (This species is also found in Ecuador and Panama.) Live specimens may be seen offered for sale on the Internet from

Megaloblatta blaberoides: Costa Rica, Guayacan 550m, Limón Province. This adult was found on a fallen tree branch in a slow-flowing stream at night. There is another roach in the photo, probably *Pseudophyllodromia* sp. © Gil Wizen

Megaloblatta on an extremely huge tree trunk in the Costa Rican rainforest © Martin Rosenberg

Nauphoeta cinerea

Opisthoplatia orientalis nymph and adult female

time to time but never with accompanying photographic documentation or facts. I observed *Megaloblatta* specimens brought back under permit and reared in a lab in 1995, but the survival rate was low and subsequent giant oothecae did not hatch. This genus is associated with large trees in the rainforest. Egg laying specifics are unknown, but they are likely deposited under loose bark. Nymphs are dark chocolate brown with reddish legs and a pair of orange spots on the abdomen. There is also a visible milky secretion on the last tergite (like bullet roaches or *Eurycotis lixa* at times). In addition to warning coloration, the nymphs make squeaky noises by shaking the abdomen to startle predators. Adults are very dark red-brown, appearing black from far away, without any contrasting coloration.

Despite many competing claims stating modern and prehistoric cockroaches are as large or larger, documentation is always lacking. The longest independently confirmed cockroach in the world is a 97 mm, dried specimen of *M. longipennis* from Peru. This record has remained on the Guinness World Record list for more than twenty years. The few dead specimens I have seen are in the upper size range of *Blaberus giganteus*. Like *Blaberus*, the foot lacks an arolium (the pad between the claws) that helps many other roaches to climb glass. A similar looking species that averages slightly smaller is called *Megaloblatta blaberoides*. Its species name comes from this genus' strange similarities to the *Blaberus* living in the same rainforests.

Nauphoeta cinerea (Olivier, 1789)
Lobster Cockroach, ACS #51
Adult length: 23-26 mm
Glass climbing: Yes
Reproduction: Live-bearing
Although it has been commonly offered since only the late 1990s for feeding reptiles and inverts, *N. cinerea* has been maintained in laboratory cultures since at least the 1950s (Roth and Willis 1957b). The common stock kept in the United States was first collected from the "wild" in Florida in the late 1980s. It originated in Africa. As a result of human activity the lobster cockroach has long been found throughout the world in tropical areas, as well as greenhouses in cooler areas. Though widespread, it is not a commonly encountered species so there is a reasonable chance all stock kept in the United States is descended from the original Florida culture. This species is listed in more comprehensive pest control identification guides as an occasional pest of greenhouses and is reportedly said to feed on *Panchlora nivea* found in the same greenhouses. The occasional occurrence in greenhouses is likely, but the predation report is highly dubious (McMonigle 2009d).

The lobster roach is reared commonly as a food for smaller herps and inverts. While some people believe the name 'lobster' results from its epicurean taste to predators, the name comes from markings on the pronotum that someone thought looked like a lobster claw (this takes some imagination). The species name *cinerea* is from the Latin word 'ciner' meaning ashes—it most certainly refers to the gray color of the adult's wings. The nymphs are brown. This is the only species in the genus *Nauphoeta*.

The live-bearing *N. cinerea* female produces about two dozen young after forty days of gestation. Nymphs grow quickly to adulthood in just a few months; adults live up to a year. The male looks similar to the female but is slightly smaller. They can be sexed early as nymphs by differences in the ventral abdominal segmentation. Both genders have wings about as long as the abdomen. Reproduction

Nauphoeta feeders

Neostylopyga rhombifolia

and growth are affected by temperature more than food.

All stages can climb glass, so they are generally offered to predators in small cages or from feeding dishes with the upper sides coated with petroleum jelly to prevent escape. Newborn nymphs are many times larger than a fruit fly, but are valuable food for any tarantula spiderling. Adults are rarely more than twice the mass of a feeder cricket, which limits their use as food for large reptiles.

Neostylopyga rhombifolia (Stoll, 1813)
Harlequin Cockroach, ACS #52
Adult length: 20-25 mm
Glass climbing: Yes
Reproduction: Egg-laying
The harlequin was considered one the prettiest cockroaches (McMonigle and Willis 2000) before availability of *Therea* and others. Different stocks can vary in the intensity of the dark coloration—the darkest stocks lose much of the abdominal patterning into one big black splotch. The species name *rhombifolia* may refer to the pattern on the abdomen (only visible on light-colored forms). Adults have only vestigial forewings, but make up for their lack of flight with their incredible land speed. They eat any common cockroach fare and are similar in care to other blattids such as *Periplaneta*. Nymphs grow to maturity in about four months, while adults live close to a year, rarely longer.

Neostylopyga rhombifolia is probably from southeast Asia, but has been introduced to Mexico and many tropical areas. This species has been reared in North American labs for quite some time (Roth and Willis 1960). Outdoors, a specimen was collected from Nogales, Arizona, in the early 1900s (Helfer 1963). I performed limited scouting around that city in 2009, which turned up none; no pictures of Nogales specimens show up in web searches. This cockroach may be found in any tropical area of the world due to transport by man. One, dark-colored, common stock in the 1990s was collected from an introduced population in Hawaii. Many culture stocks have come and gone over the decades. Specimens are commonly acquired that refuse to lay eggs regardless of the food, temperature, and moisture provided. However, loss of cultures is probably more often due to their high reproductive abilities in captivity that result in massive numbers, followed by overpopulation crashes or reduced interest on the part of the hobbyist.

Opisthoplatia orientalis (Burmeister, 1838)
Red and Black Cockroach, ACS #53
Adult length: 26-43 mm
Glass climbing: Yes
Reproduction: Live-bearing
This brightly colored cockroach was commonly available around 1996 (stock from Japan) and has become a favorite pet (McMonigle 2004b). Nymphs are light tan, with textured markings that make them look like a small wood chip. Nymphs blend in well with dead leaves and weathered wood. Tan coloration grows darker and is still grainy and wood-like to subadult. The body changes little in shape between hatchling and adult, but the appearance of the exoskeleton radically transforms. Upon the molt to adulthood, the animal transforms to a distinctive glossy black, with broad candy apple-red margins, red legs, and a white margin across the front of the pronotum. Both genders have the same color pattern, though the male is more intense. Adults possess very tiny wingpads that are only visible from certain angles.

The red and black cockroach is not an overly energetic species. Nymphs and adults normally stay near the bottom of the cage,

Neostylopyga rhombifolia, La Reunion Island (PD) Philippe Vieux-Jeanton

Opisthoplatia orientalis nymph © Alex Yelich

Opisthoplatia orientalis © Alex Yelich

though they can climb nearly any surface. They do not burrow in the substrate and prefer to hang out on wooden decorations. Egg carton can be used in place of wood, but it has to be replaced often since the roaches chew it up. *Opisthoplatia* can climb smooth surfaces, so petroleum jelly should be placed around the upper lip of the cage. Although a member of the Epilamprinae, which contains many moisture-sensitive, semi-aquatic species, this species can survive a bit of dryness. Specimens swim readily if tossed in an aquarium, but do not appear to dive. Although it has long been considered an aquatic species (Takahashi 1926), this detail is often overlooked.

In the case of *O. orientalis* it is easy to pick out adult pairs with 100% accuracy using size. Females are always a minimum of twice the mass of males. Females normally give birth to 30-40 young at a time. The species is not always easy to culture and females can die of old age without ever giving birth to young. Following maturity, it is normal for months to go by without any sign of babies. Following a slight rise in temperature, a few rose leaves, and some patience, overnight the cage is filled with an unbelievable number of tiny wood chips (hatchlings). Hide-spot preference, disposition, and diet are similar to hissers.

Scientific names often offer unique descriptions or insight into the animal. The scientific name for the red and black cockroach does not convey much detail. There is no mention of the adult's spectacular colors or the nymph's wood-like appearance. *Opistho* translates as 'rear' or 'abdomen' and *platia* means 'flat.' The abdomen is dorsally flattened, but not much more than any other cockroach. *Orientalis* means 'eastern' and is a general reference to where the species lives. Its home includes tropical forests in Japan and China.

Oxyhaloa deusta (1784, Thunberg)
Cape Redhead Cockroach, ACS #97
Adult length: 12-21 mm
Glass climbing: Yes
Reproduction: Live-bearing

When I first learned there was a member of the *Oxyhaloa* genus in captivity I was excited. The only species I knew about was a metallic-green and orange buprestid beetle mimic (*Oxyhaloa buprestoides*). It was disappointing to learn this new species did not mimic a metallic jewel beetle, but the overall look and even the colors were still quite nice. The Cape redhead cockroach (*Oxyhaloa deusta*) is nicely marked in satiny black and has a brick red head (it shows up bright red with flash photography). Other cockroaches with older common names like death's head and orange head refer to the animal's pronotum coloration, whereas in this species the reference is the actual head. The red head and black body seem to be part of a mimicry complex involving redheaded *Bromophila* flies, *Deropeltis* spp. nymphs, and *Mylabris* blister beetles. This species is capable of producing a defensive odor that smells a little like crushed garlic.

From the photographs I was expecting something about the size of *Nauphoeta cinerea*, but again I was jumping the gun. I was surprised to discover an adult female (the larger of the two genders) was barely more than half an inch.

This sturdy species is a member of the subfamily Oxyhaloinae along with the hissing cockroaches, lobster roaches, giant lobsters, etc. All stages are able to climb smooth surfaces readily, but they usually stay on pieces of wood in the cage and avoid climbing. It breeds quickly for some people, but for others, mothers never produce young (McMonigle 2011d). This may relate to the number of specimens started out with. If started with two dozen or

Oxyhaloa duesta nymphs and adult female

Oxyhaloa duesta adult male

Panchlora nivea adult females (males are notably smaller)

more founders, colonies seem to be easily maintained. Nymphs are hardy and grow readily on a diet of dog food and fruit.

Panchlora nivea (Linnaeus, 1758)
Green Banana Cockroach, ACS #56
Adult length: 12-23
Glass climbing: Adults
Reproduction: Live-bearing
The green banana roach is a beautiful light green with a yellow stripe running along each side of the pronotum and fore tegmina. It is a must-have for the cockroach aficionado. This New World species is among those that have benefited from man's habit of shipping fruit, namely bananas, around the globe. *Panchlora nivea* occurs over a huge range, including Central and South America and the West Indies.

It has shown up around the world and has long been established in North America. *Panchlora nivea* was naturalized in the United States probably more than a century ago. Hebard reported it was well established in Texas in 1917. This species is found along the Gulf Coast from Florida to Texas. It has been found at California ports, but for some reason does not seem to be established west of Texas (McMonigle 2009b). Commonly kept stock dates back to before 1994 and may be directly descended from Dr. Roth's culture. "I started a culture of *Panchlora nivea* from a single pregnant female I found on bananas in a grocery store back in the 1950s. I have a culture of *nivea*—but my office is too small to keep more than the 5 species I culture here" (Roth pers. comm. 2002). Rearing this species in captivity dates back even further (Davis 1930).

This unidentified species was collected in Belize and is currently being cultured. Care is identical to *P. nivea*. Nymphs are orange-colored and very fast. Adults are excellent flyers. © Gil Wizen

A blue *Panchlora* relative from Guyana © Arthur Anker

This species only survives in warmer temperate areas, but the northern range in the USA is uncertain. In colder areas (and in other countries around the world) it has been encountered in greenhouses, likely transported with plants from Florida or the Caribbean. This species, which are strong flyers, cannot survive in homes or buildings, but may fly in from outdoor populations or reproduce in greenhouses.

Outdoors, the spectacularly cryptic, green adults can be found during the day resting between green leaves. They are capable of climbing smooth, vertical surfaces such as glass or plastic. Adult *P. nivea* are very energetic and have the ability to fly well. Silicone barriers impede escape but do not prevent it; therefore, a bucket cage is necessary to prevent escapes. Nymphs are dark brown and resemble the immatures of other genera like *Pycnoscelus*. However, all but the smallest nymphs can be identified because they are greenish on the underside (additionally, *Pycnoscelus* spp. nymphs can also climb glass). The 1st instar young of this live-bearing species are tiny at 1.5 mm, approximately the size of a fruit fly. The immatures are incapable of climbing smooth, vertical surfaces and prefer to live a subterranean life. They tunnel through dirt and do not attempt to climb smooth, vertical surfaces. The time spent from newborn to adult lasts six months and adult females live for another six to eight months. The small males live only a few months.

Panchlora nivea adult female compared to adult female *Panchlora* sp. "giant" (pale, large specimen on left)

A unique physical aspect of this species is possession of specialized barbs on the rear legs of the adults. All cockroaches have spines on the legs, but *P. nivea's* can break off and cause sharp, rapid pain. It is a rare occurrence, but if you handle adults regularly, occasionally a spine will pierce your skin and cause pain that feels like a two-second bee sting (not the feeling of a pin prick as with other species).

As with most of the popular pet roaches, *Panchlora* are live-bearing blaberids. Two to three dozen young huddle around the mother until they have changed from white to brown. They like to nibble on rotten wood, leaves, fruit, and dry dog food. Damp substrate should be provided for burrowing nymphs, while flat pieces of bark can be offered for adult hiding spots.

Panchlora sp. 'Giant'
Giant Panchlora, ACS #56
Adult length: 18-27 mm
Glass climbing: Adults
Reproduction: Live-bearing

Whenever there is a very handsome-looking but tiny cockroach like the green banana cockroach (*Panchlora nivea*), there is a tendency to hear, "Wow, that would be a really great species if it wasn't so small." As if in answer to the enthusiast, before the turn of the century, a much larger species of *Panchlora* showed up in the hobby (McMonigle 2007d). Adults are more than twice the size of *P. nivea*, are a notably lighter shade of green, and lack the yellow racing stripes. Even before the turn of the century there was no information on where the first stock originated beyond 'tropical New World.'

Panchlora sp. "giant" adult and nymph

Like *P. nivea*, males are noticeably smaller than females, but the size discrepancy is far more exaggerated. In *Panchlora* sp. 'giant,' the females at 27 mm are many times the mass of males at 18 mm (measurements taken with vernier calipers from the front of the pronotum to the end of the tegmina). Adults are excellent flyers and climbers, while the brown nymphs are earthbound and prefer to hide in the upper half inch of the substrate. A damp area in the substrate is important for the health of nymphs. Extremely tiny young, the size of small fruit flies, hatch after ova have developed inside the mother's body for a few months. *Panchlora* species tend to fare best if a partly rotten log is placed in the cage. They chew through rotten wood, but the primary diet should include dried dog food or an equivalent. Fruit, including melon rinds, pumpkin, and apple, is a better primary diet (better in terms of survival and growth in captivity), but it molds quickly and must be replaced regularly unless the culture is very large.

There are more than thirty-five *Panchlora* species found across the tropical Americas and West Indies, and the majority of species are green. I sent preserved specimens of adult males and females of *Panchlora* sp. 'giant' to Dr. Roth in 2002. He was familiar with identifying this genus (taxonomists usually specialize on certain groups or regions, so he could not help with *Gromphadorhina*). He said that without a collection location (at minimum to country) it would be impossible to determine the correct species name. "They are very similar superficially and are difficult to identify."

Panesthia angustipennis angustipennis (Illiger, 1801)
Orange-Spotted Cockroach, ACS #58
Adult length: 44-56 mm
Glass climbing: No
Reproduction: Live-bearing

Nymphs of the orange-spotted cockroach are shiny jet black with four bright orange spots. The adults of this species are fully winged (usually) and no longer have orange spots. The adult male gains a notch at the front of the pronotum, above the head, which is used to grab onto a competing male's pronotum when sparring over a mate. Males and females have similar pronota. This *Panesthia* species prefers dog food and fruit over rotten wood in captivity. Nymphs spend about two years growing slowly to adults; there is no notable behavior seen between adults and nymphs in captivity. Wing biting is possible, but the adults keep long wings and live an additional three to four years (McMonigle and Willis 2000). Like the following subspecies, low substrate moisture levels result in rapid loss of the culture even if it does not seem terribly dry. Captive

Panchlora sp. "giant" adults and a large nymph

Panesthia angustipennis cognata nymphs and adults from Cambodia © Sebastien Marcaillo

specimens maintained in the 1990s originated in the Philippines and were much larger than some other geographic populations. This subspecies ranges from India to southeast Asia. Bragg (2012) reports rearing consecutive generations of a stock of this species collected from Sabah, Borneo, in 2001. They were initially reared on decaying wood, but grew much better on fish food pellets and assorted fruit.

Panesthia angustipennis cognata Bei-Bienko, 1969, the orangeback cockroach ACS #142 is similar to the above, but adults tend to be much smaller (~30 mm). The nymphs have one large orange spot covering much of the center of the meso- and metathorax.

Panesthia angustipennis angustipennis orange-spotted nymph from the Philippines

Panesthia angustipennis spadica (Shiraki, 1906)
Japanese Wood Cockroach, ACS #57
Adult length: 36-45 mm
Glass climbing: No
Reproduction: Live-bearing
Adults and nymphs appear to feed only on rotten wood and dead leaves. *Panesthia a. spadica* digs tunnels through soft, rotten logs on the forest floor, living in social groups. Adults seem to have short wingpads, but the wings are victims of obligate wing biting (an animal would have to be kept alone to retain its wings). Nymphs spend many years reaching adulthood—adults can live up to five years (McMonigle and Willis 2000), but one forgotten watering can destroy a colony no matter how many years were invested. Bragg (1997a) states: "Nobody seems to have had much success rearing members of the Asian genus *Panesthia* . . ." Nymphs are solid black, lacking orange markings on the abdomen that vary in shape or number on other subspecies. This small-bodied subspecies is only found in Japan and Taiwan.

Paranauphoeta discoidalis (Walker, 1868)
Indonesian Assassin Mimic Cockroach, ACS #125
Adult length: 21-26 mm
Glass climbing: Yes
Reproduction: Live-bearing
Though other species in the genus *Paranauphoeta* have seen limited success in the last decade or two, *P. discoidalis* is probably a very recent addition to the cockroach hobby (McMonigle 2015b). Like other members of the genus, the adult specimen is patterned in black and white in a pattern suggestive of the warning coloration of some assassin bugs. Nymphs are black when very small, but develop a large orange splotch on the abdomen after the second molt. The subadult sports a fantastic combination of the nymph and adult color patterns, with a large orange spot on the abdomen and white markings on the pronotum and thoracic segments. While the adult is notably larger than the colorful, small *Pseudomops*, it is also a bit smaller than an adult lobster roach (*Nauphoeta cinerea*). Though its not a miniature species, its size is not nearly so impressive as its coloration. The warning colors of the adults and nymphs should endear this handsome mimic to any hobbyist.

Paranauphoeta discoidalis

Paranauphoeta formosana © Pascal Garbe

Paranauphoeta formosana © Pascal Garbe

Paranauphoeta are fast-moving, glass-climbing species that resemble *N. cinerea* in behavior, although they are not part of the same tribe or subfamily. *Paranauphoeta*, however, are likewise members of the Blaberidae (the live-bearing cockroach family). The female forms an ootheca and then retracts it back into her abdomen until the eggs are ready to hatch. The number of young for *P. discoidalis* seems to be less than a dozen per brood (often four to six), which is unusually small, but gestation is just a few months. Young can exhibit 100% survival with moderate care and take about four months to reach maturity (McMonigle 2015b). Adults then live around six months. All stages feed on fruit and dog food, doing well with damp substrate and bark hides.

Among the handful of amazing new cockroach species maintained in captivity in recent years, *P. discoidalis* is one of the prettiest and most unique. It has proven easy to culture and may one day become a useful feeder. Its current value due to introductory scarcity makes feeder use unrealistic (there are likely no massive colonies currently in existence), but it seems hardy, though even a large colony would probably never reproduce fast enough. This handsome mimic is certainly a beauty that deserves to be kept as a pet for its own appreciation.

Paranauphoeta discoidalis adults and nymphs

Paranauphoeta formosana Matsumura, 1913
Formosan Assassin Mimic Cockroach, ACS #124
Glass climbing: Yes
Reproduction: Live-bearing

This animal's name comes mostly from a superficial resemblance to the coloration of many adult assassin bugs, specifically a pair of white spots centered on black wings. (The commonly kept *Platymeris biguttatus* from Africa boasts a similar pattern.) To be fair, the model for this Taiwanese species is currently unconfirmed, and the color could possibly advertise its own distaste or defense. Still, the Formosan assassin mimic's warning coloration and shape is similar to a number of less familiar species and genera of adult assassin bugs in the subfamilies Reduviinae and Harpactorinae. (The model may well be a tropical Asian assassin bug in the genus *Acanthaspis*.) While beetle mimic cockroaches are very common, and a number of species mimic wasps, *Paranauphoeta* is the only cockroach genus currently known that would seem to mimic the warning coloration of an assassin bug.

The nymphs (like assassin bug nymphs) do not have a pair of white spots on the dorsal surface of the body. Nymphs do not have wings and do not have corresponding markings on the abdomen. However, they do have a large orange spot on the dorsal surface of the dark colored abdomen. Nymphs of certain tropical Asian *Panesthia* (specifically *P.angustipennis cognata*) have an orange rectangle on the abdomen. Whether this is a case of mimicry of another cockroach, or if these cockroaches mimic assassin nymph coloration or something else, is unknown.

Although the genus name comes from *Nauphoeta*, this genus is not related (it is not in the tribe Nauphoetini, nor even the subfamily Oxyhaloinae). *Paranauphoeta* is the only genus in the subfamily, the Paranauphoetinae. Though a singular genus and subfamily, more than a dozen species are described for this blaberid.

The handsome *P. formosana* had been whispered about for years. The first pictures of captive specimens showed up around 2006. Some European enthusiasts reported moderate success breeding them, but cultures fizzled out (McMonigle 2015a). It seems *P. formosana* may not be seen anytime soon in the North American hobby.

Paraplecta sp.
Little Kenyan Cockroach, ACS #115
Adult length: 8-12 mm
Glass climbing: Yes
Reproduction: Live-bearing
This is probably the smallest species of live-bearing cockroach ever kept in captivity. Males are usually about 8 mm (1/3″) while females are 10–12 mm (<1/2″). Hatchlings are tiny, about the size of a small fruit fly (*Drosophila melanogaster*). The roaches burrow and grow much slower than *Drosophila*, so they cannot be used realistically to feed the same predators.

Although used as a feeder for perhaps the last seven years, this animal's identity has remained a secret. For a while it was listed as Blaberidae sp. 'Kenya,' which looks almost like a scientific name with Blaberidae as the genus, but Blaberidae is not a genus, it is the name of the family of live-bearing roaches to which this species certainly belongs. Some people suggested it was *Pycnoscelus femapterus* or another species in the genus *Pycnoscelus*, but *P. femapterus* is the only *Pycnoscelus* with wingless females (Roth 1998). These females actually have small wings. This species has far fewer spines on the femora, and the short-winged males cannot fly. Male *P. femapterus* are expert flyers. *Paraplecta* may superficially resemble *Pycnoscelus*, but they are structurally different and come from a different subfamily (the Parasphaeriinae rather than Pycnoscelinae). Little Kenyans are most likely the short-winged form of *Paraplecta parva* as described (Roth 1995), but the rear wing structure would be the easiest way to verify after genitalia structure. In *P. parva,* males with short tegmina have miniature, vestigial rear wings. Both males and females are flightless, though wild stocks of this species can have fully winged, flying males. The only other similar described species is *Paraplecta minutissima*, which can resemble certain specimens due to variability in the tegmina shape and size. Identified specimens of *P. parva* are known from Guinea and Uganda, while currently maintained stock was collected in Kenya on or before 2008.

Little Kenyan roaches are very easy to rear in captivity if provided moisture and food. Fruits and dry dog food work well as a diet. Specimens of both genders can climb glass at all stages, but they usually stay in the substrate. Nymphs start out translucent yellow and progress through darker shades of tan and brown until they are nearly black as adults.

Paratemnopteryx couloniana (Saussure, 1863)
Little Goblin, ACS #100
Adult length: 16-20 mm
Glass climbing: Hardly
Reproduction: Egg-laying
This small, unassuming species is native to Victoria, Australia, and was long ago introduced

Paraplecta sp.

Paratemnopteryx couloniana female eating dog food and surrounded by nymphs

Paraplecta sp. adult female (left) vs. adult female *Pycnoscelus femapterus* (right). (*Pycnoscelus femapterus* is much smaller than other *Pycnoscelus* spp. kept in captivity.)

Paraplecta sp., Kenyan pair (male top right)

Paratemnopteryx couloniana: adult males like this one usually have longer wings.

to New Zealand, origin of the captive stock. Although it may seem to have shown up in recent years as a "new" species in captivity, stock of this animal dates back more than two decades (McMonigle 2014c). The reason for this misconception is a combination of limited popularity and moderate confusion in the identification. These were labeled '*Temnopteryx colonus*' and '*Shawella coloniana*.' *Temnopteryx* is an African genus, though some of the North American *Parcoblatta* were originally placed there.

The little goblin is relatively easy to keep and reproduce in captivity, but so far has never made a big splash as a feeder. Its ability to climb glass is very limited and it does not burrow, which makes it ideal for a variety of feeder applications. However, its small size leads to minimal use except as prey for the very early stages of small predators. It is difficult to produce large quantities, by weight, required for rearing numbers of animals. It is an egg-laying species and high humidity is preferred for all stages. Despite this, ventilation is a requirement common to all blattids. At least a square inch of screening should be part of any cage. Most fruits and pet foods (like dry dog or cat kibbles) can be used as a staple diet. The nymphs seem to eat and grow slowly. Oothecae are small and take months to hatch.

This is the only Australian species kept outside Australia proper with a long captive history. Today's stock may have come from adventive New Zealand specimens collected some time in the 1980s. The widespread cultured form is reddish, has short, slightly curled wings, and cannot fly. Wild forms usually have longer wings on males or, less commonly, are nearly black in color (Rentz 2104).

Paratropes lycoides Serville, 1838
Lycid Mimic Cockroach, ACS #136
Adult length: 25-30 mm

Paratropes sp., collected from Sucumbios, Ecuador, east of Nueva Loja © Tyler Hedlund

Glass climbing: Yes
Reproduction: Egg-laying
Paratropes are among the most beautiful cockroaches in the world. Adults mimic the coloration and shape of certain species of net-winged beetles (Coleoptera: Lycidae). They are likewise active by day and live on leaves in the canopy. "I recall seeing a specimen or two, 20-30 m above the ground in Panama back in 2001" (pers. comm. Schwarz). Very little is known of the captive habits of this group, since none have made it to culture so far, but it is probably fast growing and short-lived like other ectobiids. A number of similar *Paratropes* species are found from Mexico to northern South America.

Parcoblatta lata (Brunner, 1865)
Broad Wood Cockroach, ACS #59
Adult length: 16-25 mm
Glass climbing: Yes
Reproduction: Egg-laying
Males of the broad wood cockroach are light tan and fully winged. Females have short, rounded tegmina. The oothecae are especially sensitive to low moisture. They are deposited under loose bark. Adult males live barely a few months, while females seldom survive past four

or five months. *Parcoblatta* are generally easy species to keep in vented deli-cups as long as watering is not forgotten too often (of course, also, oothecae can mold). Still, it is easy to wipe out a colony by forgetting about it for a week or two. *Parcoblatta lata* is found across the central and eastern United States, including Florida and Ohio. Nymphs can be collected year round from inside rotten logs in sheltered forest areas and can have stripes like *Parcoblatta zebra* nymphs. They do not eat wood in captivity. In the wild, developing nymphs tend to be found in groups of six to twelve and, strangely, there often seems to be more than one species in the group (at least in northern and southeastern U.S.). The American wood roach (*Parcoblatta americana*) from California to Idaho can have similar looking adult males (or dark winged) but the females have very tiny tegmina that cover only the outer edge of the mesonotum. Boll's wood roach (*Parcoblatta bolliana*) is an eastern species also with extremely tiny female tegmina and dark colored males.

Parcoblatta lata from Montgomery, Alabama, under the bark of a standing tree © Alan Jeon

Parcoblatta pennsylvanica (De Geer, 1773)
Pennsylvania Wood Roach, ACS #60
Adult length: 18-28 mm
Glass climbing: Adults
Reproduction: Egg-laying
There are eleven other species of *Parcoblatta* found in North America; the Pennsylvania wood roach is the longest. White margins along the sides of the pronotum and the sides of the forewings combined with mottled brown arguably make this the most handsome of the *Parcoblatta*. The female's wings are usually about ¾ the length of the abdomen, but can be longer. Adult males live only a month or two at room temperature and females only three to four. Keeping this species at cooler temperatures noticeably increases the lifespan, but the nymphs may stop growing below 70° F (21° C). The Pennsylvania wood roach is found across eastern North America from Canada south to Tennessee. The southern wood roach (*Parcoblatta divisa*) is similar but smaller and with less distinct margins—it is found in Alabama, Florida, and Georgia. Caudell's wood roach (*Parcoblatta caudelli*), also from Florida, is a pale species and the only *Parcoblatta* whose females normally have full-length wings. All *Parcoblatta* males have long wings and can usually fly. They are often attracted to lights at night. In warmer areas the nymphs are as likely to be found hiding under bark on standing dead trees as within cavities in old logs on the ground.

Parcoblatta virginica (Brunner, 1865)
Virginia Wood Roach, ACS #61
Adult length: 11-16 mm

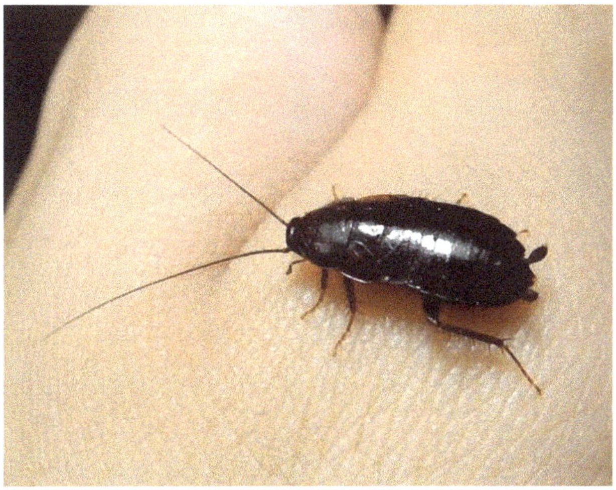
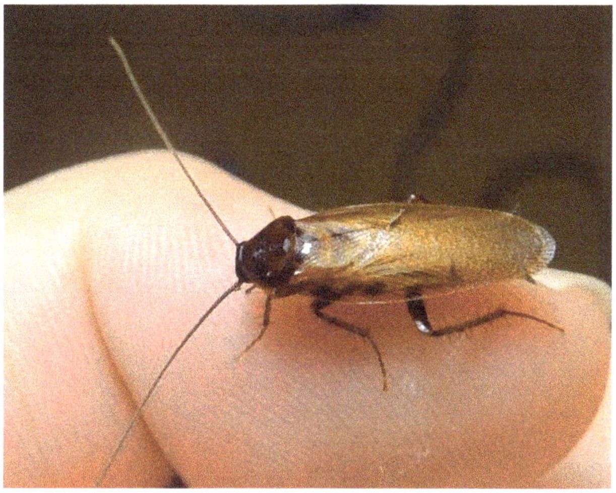

Parcoblatta americana, female (l), male (r), Chino Hills, CA © Alan Jeon

Parcoblatta bolliana, female (l) male (r), Montgomery, AL, under leaves in bamboo forest © Alan Jeon

Parcoblatta caudelli, female (l) found on trunk of oak tree, Torreya State Park, FL; male came to lights, Montgomery, AL © Alan Jeon

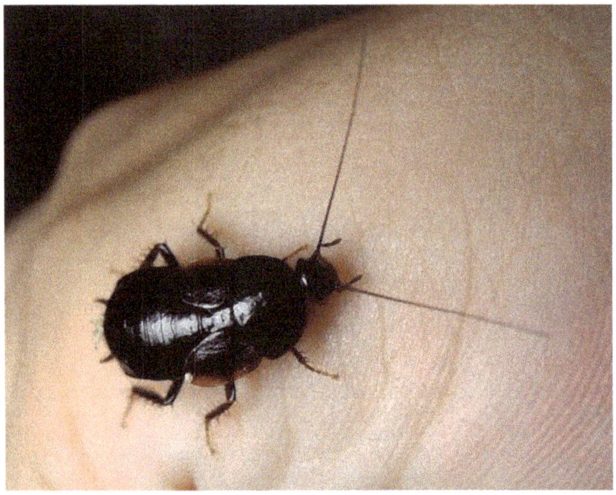

Parcoblatta uhleriana, female (l), male (r), North Carolina © Alan Jeon

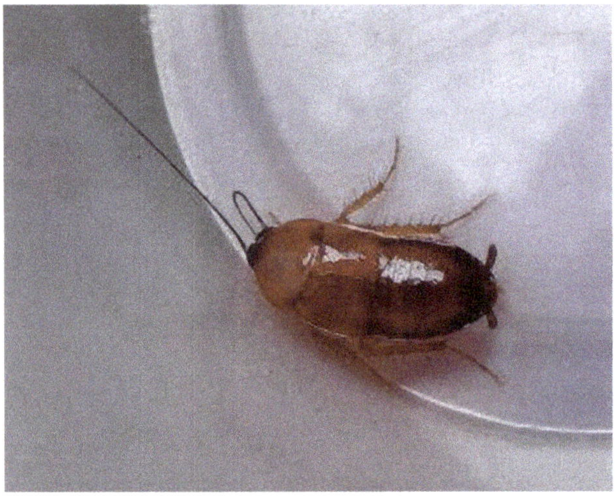

Parcoblatta virginica, female (l), male (r), collected under fallen logs, Montgomery, AL © Alan Jeon

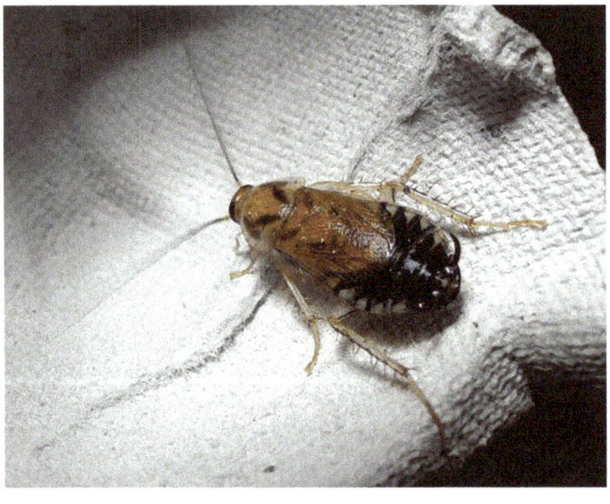
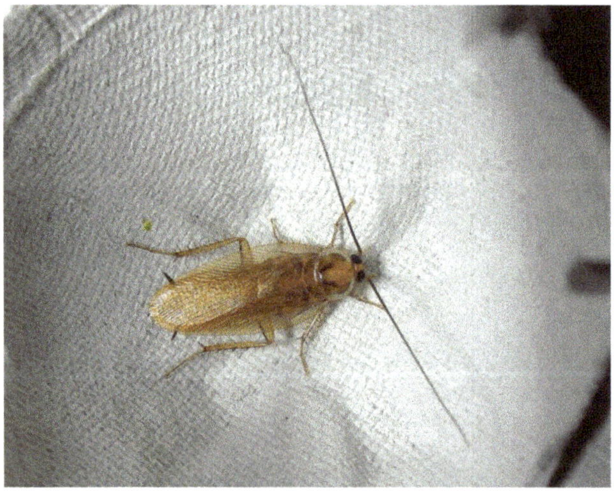

Parcoblatta zebra, female (l), male (r), from under bark of standing trees, Wetumpka, AL © Alan Jeon

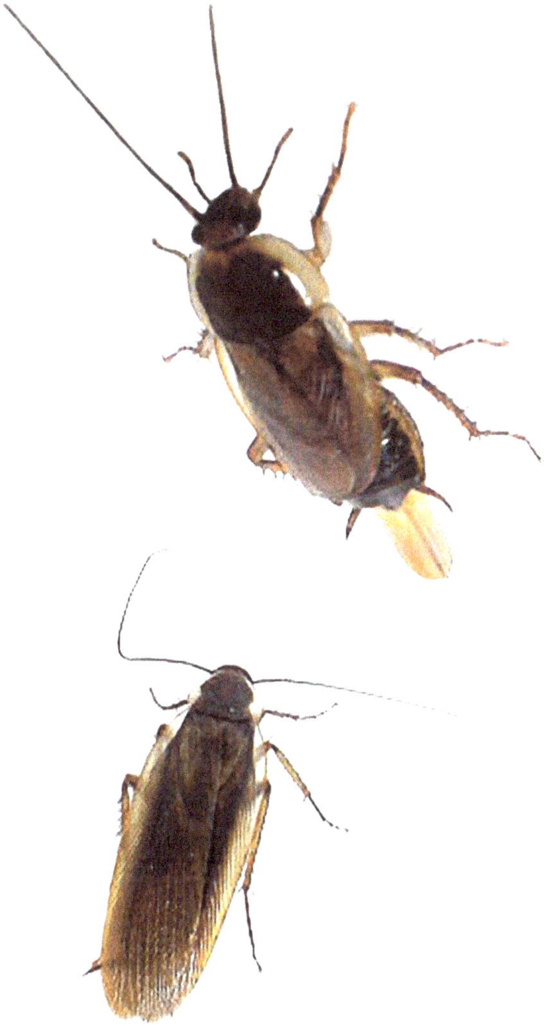

Parcoblatta pennsylvanica, adult male and female

Glass climbing: Adults
Reproduction: Egg-laying
Males of this species look like small *P. lata*. Virginia wood roach females have short tegmina and are orange-brown with a darker colored abdomen. Nymphs spend seven to twelve months reaching adulthood at room temperature. Adult males live another month or two; females may live as long as six months. *Parcoblatta virginica* is found across the eastern United States. The fulvous wood roach (*Parcoblatta fulvescens*) is similar, but the female's short wings end in a point towards the outer edge. Like most members of the genus, these species are found almost exclusively in older forests around logs and rotting stumps. The desert wood roach (*Parcoblatta desertae*) is an uncommon species from Texas found around vegetation at higher elevations, but also under large boulders in the desert.

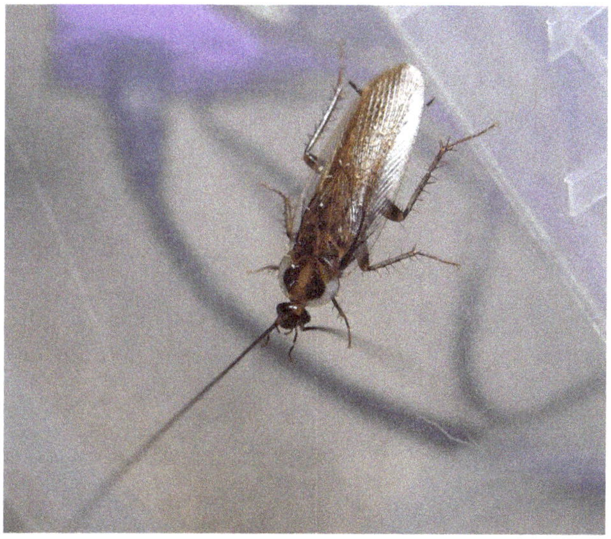

Parcoblatta divisa, male, from Montgomery, AL, collected as nymph under bark of standing trees © Alan Jeon

Parcoblatta fulvescens, female found on a side walk at night Montgomery, AL © Alan Jeon

Parcoblatta zebra Hebard, 1917
Banded Wood Roach, #62
Adult length: 17-26 mm
Glass climbing: Yes
Reproduction: Egg-laying
The banded wood roach is light tan with yellowish-white margins on the pronotum and tegmina (both male and female). The females' wings cover half of the abdomen. Abdomens are strikingly marked in dark brown and brownish-orange. This banding is, of course, most developed on the wingless older nymphs. Oothecae can be found on damp humus under loose tree bark (on large rotting timber) in mid-summer. Large nymphs can be collected from caverns within very rotten logs as soon as the ice melts in March. Such natural hides are created by other animals and decomposition, since *Parcoblatta* do not burrow or chew holes in the wood. Adult males live four or five months; females survive six months. This *Parcoblatta* is found in the central and southeastern U.S. and is not very common.

Pelmatosilpha larifuga Gurney, 1965
Vagabond Cockroach, ACS #137
Adult length: 28-35 mm
Glass climbing: Yes
Reproduction: Egg-laying
This species was described from specimens imported with fruit from the West Indies into Britain. It is a handsome striped, tropical species that does not live in human structures. The milky defensive secretion smells something like almonds. Oothecae containing 16-21 eggs take about four months to hatch. Nymphs take up to six months to mature (Mann 1995). The vagabond cockroach was on the first BCG list in 1986 and may still be in culture. It is one of the very few species with a long captive history in Europe, but nothing in North America. Another such creature is the handsome halfblack cockroach (*Brachynauphoeta brunneriana*) from Toamasina, Madagascar, stock (Hromadka 2005).

Periplaneta americana (Linnaeus, 1758)
American Cockroach, ACS #63
Adult length: 28-34 (rarely >40) mm
Glass climbing: Yes
Reproduction: Egg-laying
The American cockroach or Bombay canary is a very fast cockroach according to the literature. Despite the monikers, this species did not originate in the Americas and probably does not hail from India. It may have first come to the Americas on Spanish trading ships since an ootheca was identified from the vessel *San Antonio*, sunken off Bermuda in 1625 (Roth 1981). It may have come to Spain from Africa.

On my trip to the 2009 ACS conference in Rio Rico, Arizona, I decided to employ my superior knowledge of cockroaches and brought along a 'ringer' to the annual roach races. Thinking I was so smart, I was beaten soundly and severely by a female preteen with an adult male *Blaberus discoidalis*. *Periplaneta* land speed is usually compared to its body size or other invertebrates in scientific literature—the data does not appear to be applicable to roach racing. Nevertheless, *P. americana* is a pretty quick animal both in reproduction and in land speed. Adults fly to lights and are often found as pests around buildings in warm areas of the country. They are strong fliers, but also very clumsy, running into people now and then. This species can live entirely within buildings up north, but usually in large buildings rather than homes. I saw a single dead specimen in an old warehouse when I was a pest control operator in Ohio from 1996-1997, but a few years later I saw three live ones in the attic of an old university buildings in Cleveland (near a water purification system).

Periplaneta americana with white eyes

Periplaneta americana with white eyes

Colonies of this species emit a very strong 'roachy' smell. The ootheca usually hatches out sixteen nymphs. Adults live six to twelve months, while only three to four months are needed for hatchlings to reach adulthood. This species can be difficult to keep because the oothecae (less often nymphs) are cannibalized. Addition of cork bark that is riddled with smaller holes can protect oothecae. Laboratory rearing containers can have grating where the oothecae and smaller nymphs fall out.

Captive specimens usually do not fly, even if they are thrown into the air or dropped from a height. Wild specimens, however, can be very adept flyers. I remember watching half a dozen fly in and land on the brick wall next to outside lights at a diner near Orlando, Florida, while eating late one night in 1995. They were too high to reach to see if the group included females. In Norfolk, Virginia, 2009, I was at a front porch near midnight when an exceptionally large male flew to the light for a landing. The closest tree or house was thirty feet away and it seemed to have come from much further down the street based on trajectory.

In North America, *P. americana* may be found outdoors as far north as Virginia, where it is the only outdoor *Periplaneta*. In warmer areas like southern Georgia and Texas, *P. brunnea* and *P. fuliginosa* seem to kick out the *P. americana* in areas. The southern brown cockroach (*P. brunnea*) looks very similar to *P. americana*, but the markings on the prono-tum are indistinct and the wings are a shade darker. There is an endemic species from Australia (*Periplaneta aboriginea*), that looks identical to *P. americana*, but the tergal gland and genitalia are a little different.

Periplaneta americana 'White Eye' strain
White Eye American, ACS #64
This strain of *P. americana* has a little bit of a surprising origin story. It was not created through inbreeding in a laboratory like white-eyed fruit flies, nor radiation experiments, nor even found among a wild, tropical island population. The original specimens were found in a deep coal mine in South Wales in the United Kingdom. The white-eyed mutant form comprised about five percent of the cockroach population in the mine from 1947-1958, while the original laboratory specimens were collected around 1944 (Jefferson, 1958). The strain has been outcrossed with standard specimens in the years since its acquisition, demonstrating it is a simple recessive. It is easily isolated on the second generation (the first all look normal). Specimens with the 'white eye' gene tend to be a little more sensitive to poor husbandry than the standard form. The white eyes may be of limited use for sight, but the massive number of sensory hairs on the antennae and cerci allows the 'white eye' cockroach to appear to function normally.

Periplaneta australasiae (Fabricius, 1775)
Australian Cockroach, ACS #65
Adult length: 27-33 mm
Glass climbing: Yes
Reproduction: Egg-laying
The Australian roach probably originated in tropical Africa rather than Australia, though it is currently found almost anywhere. It is the most handsome of the *Periplaneta*, with bright yellow margins on the tegmina. Even the nymphs are brightly colored. This species is seldom found as a pest outside of the tropics and subtropics, where specimens encountered in structures have usually walked in from the outside. During a year as a pest control technician in Ohio I never encountered this species dead or alive. However, this species infested every frog, snake, mammal, and invertebrate enclosure inside two large buildings at a northern Ohio zoo with impressive voracity. (The

Periplaneta americana (USGS Bee Inventory and Monitoring Lab)

Smoky brown cockroach, *Periplaneta fuliginosa*, Chiba, Japan (CC-BY) T-Mizo

Periplaneta australasiae adults

lizard cages were mostly kept clear by their inhabitants.) This species may also be a nuisance in greenhouses. Like *P. americana*, it is easy to maintain in captivity if some attention is paid to protecting the oothecae from cannibalism. Most any pet or human food is accepted.

Periplaneta brunnea Burmeister, 1838
Southern Brown Cockroach, ACS #66
Adult length: 25-33 mm
Glass climbing: Yes
Reproduction: Egg-laying
This *Periplaneta* averages a little smaller than *P. americana* and looks very similar, but with less distinct markings. However, this species also has a brownish-black color form with light markings on the pronotum that looks nothing like *P. americana*. The nymphs have light colored markings. The adult pronotum is usually circled by a dark brown margin. It is found worldwide in warm areas. Although it is not seen in houses in cold areas, it can be a common household invader in warm temperate areas, including Alabama and Texas, USA. The southern brown and others featured here are among the *Periplaneta* species that earn the generic name, since they are found around the planet.

Periplaneta fuliginosa Serville, 1838
Smoky Brown Cockroach, ACS #67
Adult length: 26-33 mm
Glass climbing: Yes
Reproduction: Egg-laying
The smoky brown cockroach is actually shiny jet black with glossy brownish-black wings, not smoky at all. *Periplaneta fuliginosa* is not known to be a pest species in homes (at least that breeds there), but it can be a nuisance in greenhouses and outhouses in warm areas. Oothecae are hardy, but must be kept somewhat moist. Adults live close to a year. This species is found across southern North America, as well as most of Asia, South America, and Africa.

Phoetalia pallida (Brunner, 1865)
Pallid Cockroach, ACS #68
Adult length: 16-20 mm

Periplaneta japonica female © Alan Jeon

Periplaneta brunnea adult © Alan Jeon

Periplaneta brunnea nymph © Alan Jeon

Periplaneta japonica pair © Alan Jeon

Phoetalia pallida feeding.

Phoetalia pallida: small nymphs are dark brown, while large nymphs are orange.

Glass climbing: Yes
Reproduction: Live-bearing
The pallid cockroach is a small, sort of pretty species, with pink-margined adults and red-brown to brick orange nymphs. It is an excellent feeder, quick breeder, and fantastic glass climber. The adults have ash colored wings and a black pronotum like a number of Old World species, but the light margin is a delicate pink rather than white. Males and females look similar overall, but the terminal abdominal sternites are sexually dimorphic and males average smaller. The tiny segment at the end of the male is easy to pick out, as with most other members of the family Blaberidae.

It is a good feeder because it is effortless to maintain and production is high, but the adults are small, barely 4/5″ (<2 cm), and flighty. Nymphs can be handled without smashing them, but they move quickly and run up the walls of the enclosure en masse. Hatchlings grow to maturity in as little as four months, while adults live less than eight months (males less than six months). Enclosures should be well vented. Cleaning is required regularly because they do not eat their own dead. Moisture and food requirements are minimal. Any brand of dry dog food can be used and this species requires no supplementation with fruit for good reproduction.

I had used them as an occasional, if difficult to contain, feeder for years. Then one day I discovered a female in a cracker box. When used as feeders the occasional escape is to be expected, which is one reason *Blattella germanica* make terrible feeders. Usually if I discover an escaped hisser baby or other species, it is dead, wrinkled from a bad molt, or emaciated beyond survivability and usually not terribly far from the point of egress from this mortal coil. She was in a box in my cabinet, had a little cracker with her and looked fat and happy as a pig in shit. I got rid of the species immediately, never saw any more in the house for six years, and grabbed a small colony from a vendor a year or two back (McMonigle 2013a). This species is from the Florida Keys and West Indies and has been introduced throughout the tropics. Current stock comes from 1980s lab specimens of uncertain origin.

Polyphaga aegyptiaca (Linnaeus, 1758)
Egyptian Desert Cockroach, ACS #69
Adult length: 24-33 mm
Glass climbing: No
Reproduction: Egg-laying
Adult males have full-length black wings, normal rear legs, and are much smaller than the females. The rotund females might mimic big, round, black scarab beetles that are similar in size and shape. However, this association may

Polyphaga sausurrei large nymph
© Tristan Shanahan

Polyphaga aegyptiaca adult pair showcasing sexual dimorphism

Polyphaga aegyptiaca nymphs and oothecae

Polyphaga aeqyptiaca (left) and *P. sausurrei* (right) half grown nymphs

be a stretch of the imagination, and females' shells are probably covered in clay in nature rather than black. Adult females and nymphs curl up into a hemisphere and play dead when handled. Females are mostly black in color, though nymphs are varying shades and patterns of brown and light tan. The female cannot get stuck on her back because she has long back legs used like oars to flip herself over. Oothecae take three to four months to hatch; nymphs take a good year to mature. Both males and females live a year as adults, though males may seem short-lived since they easily die from shortages of water or food.

Polyphaga saussurei (Dohrn, 1888) (Saussure's Sand Cockroach ACS #70) is larger and may reach 50 mm in length, while *Polyphaga obscura* Chopard 1929 (Obscure Desert Roach ACS #141) grows up to 40 mm. These *Polyphaga* are very similar in appearance and care, but are slower and often peter out in captivity.

Polyphaga sausurrei adult female and unfertilized oothecae © Chen Zha

Pseudomops septentrionalis large nymphs and adults © Alan Jeon

Pseudomops septentrionalis ootheca

Pseudomops septentrionalis adult female

Most importantly, stocks of these species are parthenogenic only. These have been in captive culture since at least the mid-1990s, but they are difficult to culture.

Pseudomops septentrionalis Hebard, 1917
Pale-Bordered Field Cockroach, ACS #71
Adult length: 9-14 mm
Glass climbing: Yes
Reproduction: Egg-laying
This small species is one of the most spectacularly colorful roaches in the world. The black wings, along with red head and pronotum, are set off by painted white margins. The pale-bordered field cockroach is arguably the most beautiful North American species. The pretty little adults are day active (usually dusk and dawn). They are encountered on thistles and other field plants in the sunlight, so few people realize this bug is a roach. Some members of this diurnal genus are known to mimic wasps. Adult *P. septentrionalis* are commonly found on top of flowers and field plants during the day, but their behavior does not seem to mimic a wasp. Despite a large natural range and regular occurrence, wild adults are rarely encountered by hobbyists. The brown nymphs resemble *Parcoblatta* immatures. They are not day active and they always hide.

Adults only reach a maximum of 15 mm at maturity (minimum 9 mm). The male and female appear similar, but the male is a little smaller and his wings are dark brown compared to the female's deep black wings. Females produce dozens of small egg cases that are sculptured with longitudinal ridges. Oothecae look very much like polyphagid egg cases, but they are slightly curved and do not have a small flange at one end. They are dropped into cracks in the dirt and may be buried shallowly. Oothecae should be kept consistently damp. They hatch in thirty to fifty days. One female lays enough oothecae to produce more than three hundred young in a month. The hatchlings are tiny, less than a millimeter long. They feed well on dry dog food, but large numbers require very little food due to the tiny size. Growth to adulthood takes six to ten months and the adults only live one to two months, so it can take a long time to stagger a colony to produce continuous adults. Manning (1997) reports rearing adults to maturity in one month. In the wild the life cycle takes one year, with the nymphs overwintering and maturing around June. The reproductive rate is high, but it can be a difficult species to maintain because adults seem to mature and die at a staggered rate, so it can be hard to get an adult male and female together at the same time unless the colony is relatively large. This is not a messy species and even in quantity it does not foul its cage like some egg layers. Although capable of climbing smooth surfaces, individuals are always found on the substrate, not on the sides of the container, even when disturbed.

Their expansive natural range covers a third of the United States, from Texas to Florida and as far north as Tennessee and Virginia. Some stocks (primarily those kept in Europe) may date back a decade or two, but usually the ones available are no more than a few generations old.

Pycnoscelus femapterus Roth, 1998
Wingless Female Cockroach, ACS #99
Adult length: 12-18 mm
Glass climbing: Adult males
Reproduction: Live-bearing
Nymphs start out straw yellow, almost clear like termites, but gradually darken at each molt. Adulthood is reached in half a year; males die in another three to four months while females live eight or more. Cultures contain males and females. Facultative parthenogenesis

is very unlikely, if not impossible, for this species. I isolated twenty-five female subadult nymphs, which shortly matured never having been in the presence of a male. After twenty-four months, not a single nymph was produced and four of the adult females were still alive. Roth (1998) described this species from Malaysian origin captive stock collected in 1977. Museum specimens come from Burma, Java, Philippines, and Thailand. Oddly the current stock was traded as 'unidentified ex. Kenya' in the late 1990s.

Pycnoscelus nigra (Brunner, 1865)
Shadow Cockroach, ACS #72
Adult length: 20-22 mm
Glass climbing: Yes
Reproduction: Live-bearing
This species is similar to *P. surinamensis* and cultures likewise lack males, though *P. nigra* is not normally parthenogenetic. It lives in Burma, China, India, New Guinea, Philippines, Sri Lanka, and Thailand.

There is a 'dark-winged' form of *P. surinamensis* available in parthenogenetic cultures that has been traded for a few decades as *P. nigra*. Adults are similar to the normal Surinam, but are slightly smaller and the wings are black in color rather than tan. I sent some of these for identification years back: "Some clones from Central America look like your specimens" (L. M. Roth pers. comm. 2002).

Pycnoscelus nigra © Tristan Shanahan

Pycnoscelus striata (Kirby, 1903)
Ember Cockroach, ACS #98
Adult length: 16-26 mm
Glass climbing: Adults
Reproduction: Live-bearing
This is a handsome, reddish species with small males and females a little larger than *Pycnoscelus surinamensis* (this listed size range is essentially male near 16 mm and female near 26 mm). It is a relatively easy species, but on my second generation I accidentally dried the cage out and was left with only two large female nymphs. They matured and lived less than a year, but produced no young (suggesting they do not reproduce easily through parthenogenesis, if at all). This species lives in the tropical forests of Malaysia, Indonesia, and Borneo.

Pycnoscelus surinamensis (Linnaeus, 1758)
Surinam Cockroach, ACS #73
Adult length: 20-25 mm
Glass climbing: Yes
Reproduction: Live-bearing
Most cultures and most wild populations of the Surinam reproduce through parthenogenesis, with no males ever being present. The gestation period is about three months, but cultures

Pycnoscelus surinamensis adult female © Alan Jeon

Pycnoscelus femapterus adults of both genders and some smaller nymphs

Pycnoscelus striatus adult female; males look the same but are much smaller

still multiply quickly since every adult can produce young. This *Pycnoscelus* can be problematic in captivity due to their propensity to jump culture containers, along with only one being needed for reproduction. Adult females live for over a year and can produce hundreds of young. Females are able to fly. I have seen them flying to lights in the middle of the night at a Florida rest stop. This species is named after a country in northern South America. It is found worldwide and sometimes infests houseplants.

Rhabdoblatta formosana subadult nymph © Tristan Shanahan

Rhabdoblatta formosana (Shiraki, 1906)
Taiwanese Leaf Roach, ACS #129
Adult length: 23-33 mm
Glass climbing: Yes
Reproduction: Live-bearing
Full-grown animals are an olive color with mottled, feathery markings on the tegmina. These handsome cockroaches live in the jungle next to streams, blending in with the dead leaf litter that accumulates on the streambanks. The nymphs are aquatic and found under partly submerged stones. They take about a year to reach adulthood and the adults do not live very long. This species is found in Taiwan and is also found on Ishigaki and Iriomote Island in Japan (McMonigle and Willis 2000). Current stock is probably recently established (unrelated to stock from the late 1990s).

Rhabdoblatta formosana adult © Tristan Shanahan

Rhabdoblatta yayeyamana Asahina, 1967
Japanese River Roach, ACS #135
Adult length: 35-48 mm
Glass climbing: Yes
Reproduction: Live-bearing
The adult coloration is a mottled brown with a large, black central marking in the middle of the pronotum, somewhat like *Blaberus*. *Rhabdoblatta yayeyamana* are similar in rearing and habits to *R. formosana* and are also short lived as adults. This species can be found on Ishigaki island and Iriomote Island (Japan). It is one of the larger bodied, if not the largest species in the genus. (*Rhabdoblatta* includes 145 species that often look very similar.) This

Rhabdoblatta sp. in the rainforest, Panay, Philippines © Christian Schwarz

Rhyparobia maderae

species has been reared by Japanese hobbyists for a long time (McMonigle and Willis 2000), but almost never elsewhere.

Rhabdoblatta sp. ex. Sri Lanka
Adult length: 30-38 mm
Glass climbing: Yes
Reproduction: Live-bearing.
This medium-sized roach has a pattern on the outer wings that resembles feather or seashell patterns, as well as two dark lines on the pronotum. The pattern is a recurrent theme on *Rhabdoblatta* tegmina. Adults do not often live more than six months. This species was kept for a few years, but was probably lost in culture by the year 2000.

Rhyparobia maderae (Fabricius, 1781)
Madeira Cockroach, ACS #78
Adult length: 38-46 mm
Glass climbing: Yes
Reproduction: Live-bearing
Both genders have long brown wings with darker brown veins, which may help them hide among leaf litter. Males and females appear similar from above, though males average a little smaller. The defensive capabilities of this species are legendary. The defensive odor is among the most pungent and revolting—a combination of rotting corpse and burning tire. This is in addition to loud squeaking, sharp spiny legs, and frequent frass flicking.

Males make drumming noises at night by thumping the abdomen against bark to call females for mating (Kevan 1987). (This species is called the drummer in the West Indies, along with other roaches that are mistakenly attributed this species' behavior.) Females guard young long after birth and supposedly take them out foraging for food. In captivity the young do not seem to pay special attention to their mothers. The standard fares of dog food and fruits are accepted. Nymphs spend about eight months reaching adulthood. Males live another six months after that; females can live over a year (McMonigle and Willis 2000).

The name comes from the island of Madeira off the northwest coast of Africa, where it may have been transported by man. This circumtropical species hails from North Africa and has been found in the Americas, Java, Hawaii, Fiji, West Indies, and Philippines. Helfer (1963) says this species is found in the northeastern U.S., and Bell (1981) claims it is "well established indoors in New York City," but these claims are highly unlikely. The erroneous range data almost certainly comes from misreading reports from the co-founder of neuroendocrinology who used *R. maderae* in her groundbreaking work. Berta Sharrer caught cockroaches in Chicago basements for her initial work, but could not find a source after moving to New York in 1940. She happened upon some *R. maderae* that arrived with a shipment of monkeys from South America (Purpura 1998). Of course the question of where such large cockroaches would be found in a monkey shipment and why the shipment was from South America rather than Africa, is a slight curiosity. There might be some universities with Sharrer's original 1940 stock, though current material mostly hails from 1990s stock of uncertain origin.

Rhyparobia sp. ex. Malaysia
Gold Madeira Cockroach, ACS #79
Adult length: 33-40 mm
Glass climbing: Yes
Reproduction: Live-bearing
This species is like *R. maderae* in nearly every way, but is twenty percent smaller and a light tan-yellow in color. It is often labeled sp. '*goldi*' and may be listed under *lurida* or *capelloi*, a central African species. It may also have been

Rhyparobia sp. “goldi” nymph

Rhyparobia sp. “goldi” adult

Schultesia lampyridiformis

listed as another relative, *R. grandis* (Durkin 1999). Culture stock dates back to individuals collected from Malaysia in the mid-1990s.

The nymphs seem to grow a little more quickly, taking six month to reach adulthood. Both *Rhyparobia* species reproduce best if fruits, such as apples, are included in the diet (kibbled pet food alone may not be enough for reproduction).

Schultesia lampyridiformis Roth, 1973
Firefly Roach, ACS #80
Adult length: 13-20 mm
Glass climbing: Yes
Reproduction: Live-bearing
Adult *S. lampyridiformis* appear similar, but males are only about half the mass of the females. At maturity, these cockroaches resemble firefly species found in the same areas of the tropics. Nymphs, however, are marked much like the nymphs of *Blaberus* species. All current stock originates from Roth's specimens collected circa 1973 from bird's nests in Brazil.

The firefly roach has an extremely quick life cycle with the time between birth and death seldom lasting more than six months total. Females produce a few dozen tiny, dark brown babies about every forty-five days under good conditions. Captive cultures are most productive if bark or wood is included in the habitat. The wood is not a major part of the diet, as they prefer dog food, but cork bark or other materials are eaten along the surface until they must be replaced (every year or two). They are possibly the only non-burrowing species at this size that is easily handled, so they have proven

Schultesia lampyridiformis

very useful for feeding early instar mantids and amblypygids. The life cycle is rapid and cultures grow quickly enough to make them useful feeders.

Simandoa conserfariam Roth and Nasrecki 2004
Simandoan Cave Cockroach, ACS #101
Adult length: 30-44 mm
Glass climbing: Yes
Reproduction: Live-bearing

Simandoa conserfariam is closely related to the lobster roach (*Nauphoeta cinerea*) and giant lobster roach (*Henschoutedenia flexivitta*), and is placed in the tribe Nauphoetini. Roth and Nasrecki state: "It is close to *Henschoutedenia* and *Nauphoeta*, but lacks stridulating structures on the pronotum and tegmina which are present in some species of the other 2 genera." However, it does make squeaking noises (it stridulates). It shares husbandry requirements and behavioral traits with its cousins, though it is larger and more contrastingly marked. Individuals average double the mass of the giant lobster cockroach, but not much longer. Females are around 40 mm while males are a little over 30 mm. However, females with hyper-extended abdomens can measure near 50 mm. Overcrowded conditions will result in some males under 30 mm. Adults are prone to minimal wing biting like *N. cinerea*, but usually the damage is very minor. Newborn nymphs are brown but become more colorful after a few molts. Large nymphs are striped like giant lobsters, but are more brightly marked with yellow rather than lighter brown stripes.

Initially (2013c) I reported "their behavior is somewhat unusual in that older nymphs hide under objects and burrow in the upper level of the substrate with an intensity not seen in its relatives (or any other commonly kept species for that matter)." However, their behavior after reaching a high population density now seems very similar to *N. cinerea*. The nymphs do still act a little more skittish than those of their relatives, but now they stay on the surface or cling to pieces of wood in the cage. In the wild these cockroaches are only found buried—up to a foot (30 cm) deep in bat guano. "The cave was inhabited by a large colony of Egyptian fruit bats (*Rousettus aegyptiacus*). The number of bats in the colony was estimated in the range of 1000 to 1500 individuals" (Roth and Nasrecki 2004). The roaches were reportedly never seen walking around on the surface, which matches the initial behavior of the nymphs.

Simandoa feed readily on dog food but the favorite food is apple slices. Pear (including Asian) is accepted but peach, grapes, and banana are barely touched unless the animals are very hungry. This is one of the few species I feed 90% fruit (most are 90% dog food) because of its feeding preference. Nymphs take about eight months to reach adulthood, though this can be prolonged if kept at the lower end of room temperature. Adults and nymphs are capable of climbing glass though they prefer to hang onto pieces of wood. Nymphs may not

Simandoa conserfariam forming oothecae

Simandoa conserfariam

Colorful *Simandoa conserfariam* nymphs

climb the sides of the container due to their cryptic behavior, but they are quite capable of running up the sides.

This species seems rather difficult to breed, at first. The adults mature and mate readily, but then nothing happens. Four to five months later there are no signs of young and the older males can begin to die from old age. Finally about three dozen nymphs are born for each female. Each female likely produces three to four times in her life. In my experience this species is much easier to maintain than *H. flexivittata* (on par with *N. cinerea*). They even show up in other cages from time to time. It likely will become necessary to use it as a feeder or use other methods to keep the population in check.

The handsome *S. conserfariam* was discovered and saved from extinction at the same time (2002), since it is now believed to be wholly extirpated from the wild. Today it continues on only in captive populations. Specimens were described from a population found in the Simandou Mountain range in the Republic of Guinea, West Africa, around a cave system that happened to contain mass amounts of bauxite ore for use in making aluminum. Fortunately the company allowed material removal for flora and fauna surveys, as this species' existence or extinction would have otherwise never have been known (Roth and Nasrecki 2004). Not too many years ago, no such consideration would be given to habitat surrounding ore (altruism trumping economic or human survival only in fairy tales).

It is wonderful the describers of this brightly marked cockroach took an interest in

Supella longipalpa specimens (top and bottom) found in a laboratory at Oklahoma State University. © Christopher A. Varnon

Symploce pallens (top female, bottom male) © Alan Jeon

the survival of the species. It is also of great fortune that such a nice-looking species has proven rather easy to keep and is becoming a common staple in the hobby. It is one of the more impressive species to have arrived on scene in years and it is certainly neat to be able to show a curious person the "extinct" cockroaches in your collection. It is not possible or realistic to pretend we can save every natural habitat, but in some cases we can preserve captive populations of unique species for our children's children and enjoy it.

Supella longipalpa (Fabricius, 1798)
Brown-Banded Cockroach, ACS #82
Adult length: 9-12 mm
Glass climbing: Yes
Reproduction: Egg-laying
This worldwide pest is often found as a pest in college dormitories and is rarely seen in homes. The oothecae are short compared to *Blattella* and usually contain sixteen eggs. The casing or theca is usually red-brown, but sometimes green. Oothecae are not pasted with surrounding material but are usually glued to surfaces with the keel facing outward, away from the point of attachment. These probably once glued their oothecae to tree branches or tall grasses in an arid area of Africa. This species is sometimes called the furniture cockroach because its eggs are transported on furniture (whereas German roaches are transported inside of things like boxes or appliances, since oothecae are carried by the female). *Supella longipalpa* breed quickly and live only four to six months as adults.

Symploce pallens (Stephens, 1835)
Smooth Cockroach, ACS #83
Adult length: 10-14 mm
Glass climbing: Yes
Reproduction: Egg-laying
Adult *Symploce pallens* look like a small *Parcoblatta virginica*. The color and size are the same, plus the males have long wings and the females have short wings. The male is possible to distinguish from *P. virginica* because the styles are extremely asymmetrical; male styles of *Parcoblatta* are symmetrical. This species grows and breeds much more quickly than *Parcoblatta*. It is considered a pest in some tropical areas and is circumtropical. Current stock dates back at least to the mid-1990s.

Therea olegrandjeani Fritzsche and Zompro, 2008
Question Mark Cockroach, ACS #85
Adult length: 24-28 mm
Glass climbing: Adults
Reproduction: Egg-laying
The question mark is certainly among the most beautiful and memorable of all roaches. It is warningly colored in black and white and from the side the large white marking in the middle of each tegmen clearly looks like a "?" without the dot. The defensive smell is weak, so it may mimic the protective color pattern of another creature (unlike the following species, other animals with similar marking are not currently known). The adult body shape does resemble other beetle mimics.

Dead leaves are important for the substrate (decayed hardwood leaf mixes of beech, elm, oak, and sycamore work well, but they are not terribly picky). Substrate used below the leaves can be around an inch (25 mm) coconut fiber or peat. These substances are primarily inedible, but they retain moisture and can prevent pooling of water underneath. The main food, especially for large nymphs and females, should be dry dog food—they can eat quite a bit compared to the slow consumption of leaf litter. Oothecae have strong ridges along the sides and are tough and leathery. Like other

Therea olegrandjeani male

Therea olegrandjeani ootheca and nymphs

Therea olegrandjeani female

polyphagids, cannibalism of the eggs is not a concern. Unhatched oothecae result from too much or too little moisture. Temperature is a factor, but they are far more likely to desiccate if kept in the eighties. Less than twenty hatch after about four months. The tiny fuzzy brown nymphs grow into large, nearly black nymphs after around nine months. Adults live only three to four months.

The original specimens were traded as *Therea* sp. 'Grandjeani' starting in 2005—stock came from Andra Pradesh (near Gooty). They have been used as feeders by some enthusiasts, since reproduction levels are very high given enough time (Olsen 2015).

Therea petiveriana (Linnaeus 1758)
Domino Cockroach, ACS #86
Adult length: 20-32 mm
Glass climbing: Adults
Reproduction: Egg-laying
When Allpet Roaches was written, three "dream" species were included that did not exist in the hobby. One was *Macropanesthia rhinoceros*, which was just beginning to be kept around the world and is still extremely rare in the U.S.. A lady from Australia and an insect business in Japan provided the original photos in the book. The second was the beautiful lycid beetle mimic *Paratropes lycodes*, whose photo was taken by a nature photographer during a trip to Costa Rica. *Paratropes* seemed to be the most likely of the three to enter the hobby since its species are rather common from Mexico through Central and South America. To this day (2016) there is not a single reliable report of *Paratropes* ever reared in captivity. The third dream roach was an extraordinary beetle mimic that was round, and black with white spots. A friend supplied a dead specimen of this very impressive Chrysomelid beetle mimic he had acquired years earlier. It did not seem like this roach would ever make it to the hobby since *Poecilotheria* seem to be the only inverts to ever leave India. And yet it already had—current stock of the domino cockroach began to show up in collections just prior to the turn of the century (Durkin 1998a).

Therea petiveriana is a member of the family Polyphagidae (collectively known as desert or sand roaches) of which there are only a hand-

Therea petiveriana

Therea petiveriana with defective spots (2008)

Therea petiveriana

Therea petiveriana with defective spots (2015)

ful of species kept in captivity. This was the first that hated dryness. Current stock is slightly different than the female specimen depicted in the Allpet Roaches book. The spots on the cultured variety are more square than round, and adults average larger (still there is incredible variability in adult size and spot shape within a single inbred culture). Of course the variability has been well documented for a few hundred years in museum collections, but someone still decided to mistakenly rename the species '*bernhardti*' based on minor spot and rear wing shape in 2009. The rear wings in *Therea* are tiny, vestigial stumps.

Despite the usual habitat and common name, members of the family do not survive many days without available moisture. Polyphagids have minimal appetites and prefer rotten leaves to fruit or dog food. The roaches, adults primarily, also feed on banana, melon, grapes, peas, fresh sweet corn, and dog or cat food. One unique characteristic of this family is females normally bury the leathery, sculptured oothecae. In the close quarters of captivity, burrowing by adults and nymphs will bring some oothecae to the surface. Each female forms approximately 25 oothecae, each containing a dozen or so eggs. Nymphs are slow to hatch and slow growing, another common feature of polyphagids. They require very little attention but it can take many years to get a

Therea regularis female (left) and male (right)

Therea regularis female

culture really producing. *Therea petiveriana* take three to five months to hatch and can take more than 18 months to reach adulthood. At high temperatures, and pristine conditions, the entire cycle including gestation may take as little as 12 months (or more than 24).

This species seems to develop a disproportionately high number of males, which often outnumber females by three to one. Males are easy to identify from above by the much longer antennae. Females are short lived compared to other polyphagids and seldom continue six months after maturity. Males become adults first and rarely survive more than three months. Nymphs cannot climb smooth vertical surfaces and adult females usually do not, but adult males climb nearly anything.

The domino cockroach is thought to mimic a brightly colored tortoise beetle (family Chrysomelidae) due to its shape, but a model has not been identified. There are a number of large black arthropods with white spots from India which may comprise a complex of protected species. There are at least two large beetles—round flower scarabs (*Protaetia alboguttata*) and elongate tiger beetles (*Anthia sexguttata*). There is even a large black whip-spider with white spots (*Phrynichus phipsoni*) from Tamil Nadu, India (McMonigle 2013d).

Therea regularis Grandcolas, 1993
Orange-Spotted Domino Cockroach, ACS #105
Adult length: 20-23 mm
Glass climbing: Somewhat
Reproduction: Egg-laying
This domino roach is very similar to the common domino cockroach and the question mark cockroach, though the color of the spots is different and the adults are never very big. Instead of white markings on the adult's wings, the spots are orange (depending on who is looking at them). The color can be more of a dirty tan, sort of yellow. The nymphs are fuzzy brown, whereas the other two species have black nymphs.

The heavy-bodied female forms six to ten oothecae that it deposits in damp, loose substrate. Like other *Therea* the nymphs cannot climb, while the adults can rapidly walk up smooth, vertical surfaces. Care is standard: a few inches of substrate that is never dry and never soggy. Some old hardwood leaves and the occasional dry dog food can be thrown on top. Oothecae hatch after about six months, nymphs seldom reach adulthood in under a year, and adults live two to four months with males dying first (McMonigle 2014a).

Males can climb glass, while nymphs cannot, and adult females not really. This is the most recently cultured species available from the genus *Therea*. It showed up in culture around 2012. Specimens have been found in Gingee and Madras, India, within the natural range of *T. petiveriana* (Olsen 2015).

Unidentified ex. Tanzania
Bullet Roach, ACS #88
Adult length: 12-17 mm
Glass climbing: Yes
Reproduction: Egg-laying
Bullet roaches are nicely colored animals that require less attention than the vast majority of egg layers. Females glue the oothecae to objects but seldom, if ever, glue materials over the surface. As usual for egg layers that paste oothecae to objects, a piece of weathered or rotten wood in the cage serves as an excellent spot for eggs to develop. The current stock comes from a handful of individuals collected under a log in the late 1990s at the base of Mount Kilimanjaro in Tanzania, Africa.

Bullet roaches

Formosan sand roach male eyes

Formosan sand roach pair

Little penguin roaches

Formosan sand roach female with oothecae

Unidentified ex. Taiwan
Formosan Sand Roach, ACS #89
Adult length 18-24 mm
Glass climbing: No
Reproduction: Egg-laying
This species is similar in appearance to North American *Arenivaga*, but is much larger and more easily cultured (though small compared to the African *Polyphaga* spp.). Nymphs reach adulthood in about nine months, which is fast for a polyphagid, especially since that includes the six to eight weeks needed for the eggs to hatch. Oothecae are normal for this family—having a thick ridged, brown casing. Stock came from R. Willis circa 1997 and originated from Taiwan.

Unidentified ex. Malaysia
Little Penguin, ACS #131
Adult length 9-11 mm
Glass climbing: Yes
Reproduction: Egg Carrying
These are incredibly fast-breeding, rapidly growing, miniature cockroaches. They would make poor feeders because they easily damaged from handling, run fast, scale petroleum with some skill, and dry out quickly. Petroleum jelly has to be reapplied as often as a few times a week to keep them in check. They may eat it.

Like *Blattella* spp., females carry the oothecae partly withdrawn into the abdomen as the eggs inside develop. One ootheca contains approximately twenty-four eggs. When the eggs are mature, and conditions favorable, the female drops the 4 mm long ootheca. The nymphs start swallowing air and within about five minutes they begin to emerge from the ootheca. (If oothecae are artificially removed late in development, hatching begins within minutes.)

Forest cockroach, unidentified species, Kri Island, West Papua, Indonesia G. Bohne

Drymaplaneta semivittata, from Western Australia © Jean and Fred Hort

Chorisoneura texensis nymphs © Alan Jeon

Ellipsidion australe, Australia (CC-BY) Cyron Ray Macey

Ellipsidion humerale, Australia (CC-BY) James Niland

CONCLUSION

結論

Cockroaches have been around for more than 360 million years. Over 4,600 species have evolved into various shapes and colors. While many roaches have adapted to live on flowers, inside rotten logs, in deserts, and a multitude of other environments, only a handful are able to survive in our homes (without help). Many of the non-pest species make wonderful pets or even great pet food. This text hopefully advances knowledge and love of cockroaches.

Forest cockroach from the Cardamom Mts., Cambodia © Arthur Anker

A colorful cockroach from S. Bahia, Brazil © Arthur Anker

Cockroach (Blaberidae), Sarawak, Malaysia (CC-SA) Bernard Dupont

GLOSSARY 專業詞彙

Abdomen: Rear section of the insect body, attached to the thorax; contains the genital organs and cerci.

Adventive: Non-native species that has become established; naturalized.

Allometry: The disproportionate symmetry associated with armature of the pronotum or legs relative to total size and usually associated with male secondary sexual features.

Apterous: To be without or lack wings.

Autotomy: The reflexive release of an appendage at a specific regrowth point.

Arolium: The terminal pad between the claws of the tarsus.

Blatticulturist: Person who keeps cockroaches in captivity.

Cercus (*pl.* Cerci): Paired appendages at the end of the insect's abdomen.

Colleterial Gland: Paired female organs that produce the material covering the ootheca.

Coxa (*pl.* Coxae): Basal leg segment.

Crypsis: Mimicry of common background objects like rocks or plants rather than fauna.

Dentate: Having a series of small ridges; toothed.

Desiccate: To dry out.

Dimorphic: Possessing two distinct forms or shapes. Sexually dimorphic males and females are visibly different in body form.

Dorsal: The upper surface or top.

Endemic: Native to a geographic area.

Euplantula (*pl.* Euplantulae): Adhesive pad on the cockroach tarsus used in climbing smooth surfaces; similar to pulvillus.

Eusociality: The highest level of social organization in animal society characterized by division of labor and cooperative brood care.

Extant: Currently living species.

Exuvia (*pl.* Exuviae): The shed exoskeleton.

Femur (*pl.* Femora): Thigh; segment of an insect leg between the trochanter and tibia.

Frass: The solid waste of invertebrates.

Holotype: A single physical example or drawing of an organism used in the description of a species or subspecies.

Hypopus: A unique stage of certain types of mites that is a response to reduced food, where they molt into a suction cup stage and search out exoskeletons of living arthropods to cling to; phoretic stage.

Instar: Nymphal stages after hatching and between molts (i.e. a third instar nymph would have molted twice). Common abbreviation for instars is L1, L2, etc. The L comes from *Larven-stadium*, the German word for instar.

Intersegmental Membrane: Elastic, less sclerotized portion of the exoskeleton joining segments.

Larva (*pl.* Larvae): The stage between egg and pupa; usually reserved for insects with a complete metamorphosis.

Margin: A border or edge.

Mandibles: Central chewing pair of mouthparts; jaws.

Maxillae: Paired mouthparts located behind and outside of the mandibles—basal segments usually end in a sharp point and function like pincers, while the extremities form maxillary palps.

Mesonotum: Upper shield or plate on the middle section of the thorax; covering the mesothorax that bears the tegmina and middle legs.

Metanotum: Upper shield or plate on the rear section of the thorax; covering the metathorax that holds the rear wings and hind legs.

Molt: The act of shedding the exoskeleton.

Notum: The upper segments of the thorax.

Nymph: Stages between egg and adult for insects with incomplete metamorphosis.

Obligate Symbiotic: Refers to one creature that cannot live without the other under normal circumstances. In the case of certain mites, the mites cannot survive without the cockroaches, but the cockroaches can survive without the mites.

Ocellus (*pl.* Ocelli): Light detection organs.

Ommatidium (*pl.* Ommatidia): Structural unit or complete facet of the compound eye.

Ootheca (*pl.* Oothecae): A group of eggs produced together and surrounded by a hardened outer shell; egg case.

Ovipositor: A tubular structure through which a female lays eggs.

Palp: Short for palpus; elongate, segmented appendage normally used to taste and manipulate food and located near the mouth of an arthropod.

Parthenogenesis: Reproduction through the development of unfertilized eggs.

Phoretic: Describing an animal which attaches to another animal for transportation.

Profemur (*pl.* Profemora): Third leg segment (femur) of the front pair of legs attached to the prothorax.

Pronotum: Shield-like upper surface of the prothorax.

Prothorax: Front section of the thorax; attached to the head and contains front pair of legs.

Sclerotize: To cause rigidity of the exoskeleton through addition of the protein sclerotin or a substance other than chitin.

Seta (*pl.* Setae): Hollow growth from the epidermal layer of the body wall that is often sensory in function; hair-like setae are often referred to as 'hair' but are principally different from the hair of mammals.

Sexually Dichromatic: Of or having differently colored male and female forms.

Sexually Dimorphic: Presence of noticeably different external gender characteristics.

Spermatheca: Female's sperm storage compartment.

Spermatophore: A specialized sperm packet.

Spiracle: The external opening of the tracheal respiratory system; breathing hole.

Sternite: A thickened ventral plate covering a bottom segment of the cockroach body.

Stridulation: The act of making noise by rubbing body parts together.

Tarsus (*pl.* Tarsi): The foot of an insect. The tarsus contains a number of small segments including the claws.

Tegmen (*pl.* Tegmina): The tough and thickened front pair of wings.

Teneral: Soft, often pale or white-colored state of the insect exoskeleton immediately following a molt, before the cuticle has darkened and hardened.

Tergite (sing. noun): A thickened dorsal plate covering an upper segment of the cockroach body. (Tergal, adj.) of the tergite.

Orange-spotted cockroach, *Sundablatta sexpunctata*, Sarawak, Malaysia (CC-SA) Bernard Dupont

Pseudophoraspis sp., Perak, Malaysia (CC-SA) Bernard Dupont

Methana cockroach, *Methana marginalis*, Brisbane, Queensland (CC-BY) James Niland

A day-active cockroach, *Phoraspis* sp.—and possible tortoise beetle mimic, Brazil © Arthur Anker

Theca: The proteinacious (often thick and leathery) shell that surrounds the eggs.

Thigmotactic: A preference to feel surfaces against the top and bottom of the body.

Thorax: Middle section of the insect's body; in between the head and abdomen.

Tibia (*pl.* Tibiae): Segment of an insect leg between the tarsus and femur.

Trachea: Internal breathing tube that opens at a spiracle and branches out to the body cells.

Trochanter: Leg segment between the coxa and femur.

Ultimate Molt: The final shedding of the exoskeleton from an immature nymph resulting in a sexually mature specimen.

Ventral: Bottom or underside.

Volvating: The property of being able to curl into a spherical shape.

Cockroach from the Kelabit highlands, Sarawak, Malaysia (CC-SA) Pavel Kirillov

Cockroach from Sarawak, Malaysia (CC-SA) Pavel Kirillov

Parental care in cockroach (*Pseudophoraspis* sp.), Sarawak, Malaysia (CC-SA) Pavel Kirillov

Cockroach from Sarawak, Malaysia (CC-SA) Pavel Kirillov

Balta denticauda, Queensland, Australia (CC-BY) Donald Hobern

Laxta granicollis adult male. Adult females resemble a *Blaberus* nymph. This species is known as the common trilobite roach in Australia. (CC-BY) Donald Hobern

BIBLIOGRAPHY 參考書目

Arnett, Ross H., Jr. and Richard Jacques, Jr. (1981) *Simon and Schuster's Guide to Insects*. Simon and Schuster, New York.

Banks, W.M. (1969) Observations on the rearing and maintenance of *Blaberus giganteus* (Orthoptera: Blaberidae). *Annals of the Entomology Society of America* 62(6): 1211-1312.

Barth, Robert H., Jr. (1964) The mating behavior of *Byrsotria fumigata* (Guerin) (Blattidae, Blaberinae). *Behavior* 23: 1-30.

Barth, Robert H., Jr. (1968) The mating behavior of *Gromphadorhina Portentosa* (Schaum) (Blattaria, Blaberoidea, Blaberidae, Oxyhaloinae): an anomalous pattern for a cockroach. *Psyche* 75(2): 124-131.

Beccaloni, George. (1987) Species Report No. 1: *Gromphadorhina portentosa* (Schaum). *Blattodea Culture Group Newsletter* 2(2): 7.

Beccaloni, George (1990) The world's largest roach. *Blattodea Culture Group Newsletter* 5(1): 4-6.

Bell, William J. (1981) *The Laboratory Cockroach: Experiments in Cockroach Anatomy, Physiology and Behavior*. Chapman and Hall, London.

Bell, William J., Louis M. Roth, and Christine A. Nalepa. (2007) *Cockroaches: Ecology, Behavior, and Natural History*. Johns Hopkins University Press, Baltimore, MD.

Bohn, Horst, et al. (2010) A jumping cockroach from South Africa, *Saltoblattella montistabularis*, gen. nov., spec. nov. (Blattodea: Blattellidae). *Arthropod Systematics and Phylogeny* 68(1): 53-69.

Bragg, Phil E. (1997a) *An Introduction to Rearing Cockroaches*. P.E. Bragg, Derbyshire, UK.

Bragg, Phil E. (1997b) Rearing aquatic cockroaches. *Bulletin of the Amateur Entomologists' Society* 56: 117-121.

Bragg, Phil E. (2012) Rearing *Panesthia angustipennis angustipennis* (Illiger, 1801) from Borneo (Insecta: Blattodea: Blaberidae: Panesthiinae). *Cockroach Studies* 3: 5-8.

Breul, Martin, and Simone Nake. (2011) *Grundwissen Schaben: Biologie - Haltung - Zucht*. Sungaya-Verlag.

Brossut, R. (1983) Allomonal secretions in cockroaches. *Journal of Chemical Ecology* 9: 143-158.

Brotherton, Bill (1986) *Basalt Bill*. Self-published, North Queensland, Australia.

Burgess, N. R. H., and K. N. Chetwyn. (1979) Cockroaches and the hospital environment. *Nursing Times* 6: 5-7.

Carpenter, F. M. (1992). *Treatise on Invertebrate Paleontology*. Part R, Arthropoda 4, Vol. 3: Superclass Hexapoda. The Geological Society of America and the University of Kansas, Boulder and Lawrence.

Carrel, James E., and Elisabeth M. Tanner. (2002) Sex-specific food preferences in the Madagascar hissing cockroach *Gromphadorhina portentosa* (Dictyoptera: Blaberidae). *Journal of Insect Behavior* 15(5): 707-714.

Chopard, L. (1938) *La Biologie des Orthoptères*. Lechevalier, Paris

Connell, Kathleen B. (2011) Dietary considerations in the reproduction of *Nauphoeta cinerea*. *American Cockroach Society Publication* 1: 2-9.

Cochran D. G., J. M. Grayson, and A. B. Gurney. (1975) *Cockroaches: Biology and Control*. W.H.O/V.B.C. Issue 75, Vol. 576, pp. 1-48.

Davis, W. T. (1930) Rearing the young of the viviparous cockroach, *Panchlora cubensis*. *Journal of the New York Entomological Society* 38(2): 85-88.

Day, M. F. (1950) The histology of a very large insect, *Macropanesthia rhinoceros* Sauss. (Blattidae). *Australian Journal of Science Res., B*, 3: 61-75.

Durkin, Adrian. (1996) Species Report: *Ergaula capucina* (Bruner v. W. 1893). *Blattodea Culture Group Newsletter* 10: 8-10.

Durkin, Adrian (1998a) New Species. *Blattodea Culture Group Newsletter* 12(1): 3.

Durkin, Adrian (1998b) New Species. *Blattodea Culture Group Newsletter* 12(2): 5-6.

Durkin, Adrian (1999) New Species. *Blattodea Culture Group Newsletter* 13(2): 7-8.

Easterday, Cary R. (2001) Fossil terrestrial arthropods of the '7-11 Mine' (Desmoinesian-Missourian: Pennsylvanian: Carboniferous), Columbiana County, Eastern Ohio. *Session No. 112. Department of Geological Sciences*. The Ohio State University, Columbus, OH.

Eisner, Thomas. (1958) Spray mechanism of the cockroach *Diploptera punctata*. *Science* (18 July).

Eisner, Thomas. (2003) *For Love of Insects*. Belknap Press, Cambridge, MA.

Field, Gordon, Linnaeus, B. Savage, and Ralph J. Duplessis. (1966) Note on the cockroach mite *Pimeliaphilus cunliffei* (Acarina: Pterygosomidae) infesting Oriental, German, and American cockroaches. *Journal of Econ. Entomology* 59(6): 1532.

Fisk, F. W. (1951) Use of a specific mite control in roach and mouse cultures. *Journal of Econ. Entomology* 44: 1016.

Fotedar, R., Shriniwas U. Banerjee, and A. Vermas. (1991) Cockroaches (*Blattella germanica*) as carriers of microorganisms of medical importance in hospitals. *Epidemiol. Infect*. 107: 181-187.

Fritzsche, Ingo. (2008) *Lucihormetica grossei* n. sp. -eine neue Art der "Leuchtschaben" aus Venezuela (Dictyoptera: Blaberinae: Brachycolini). *Arthropoda* NR.

Fujita, M. S., S. M. Himizu, and R. Achida. (2011) Establishing a culture of *Eucorydia yasumatsui* Asahina (Insecta: Blattaria, Polyphagidae). *Proceedings of the Arthropodan Embryological Society of Japan* 46: 1-3.

Ge, G. F., et al. (2012) Anti-tumor effects and chemical compositions of *Eupolyphaga sinensis* Walker ethanol extract. *Journal of Ethnopharmacology* 141(1): 178-82.

Gordon, David G. (1996) *The Compleat Cockroach: A Comprehensive Guide to the Most Despised (And Least Understood) Creature on Earth*. Ten Speed Press, Berkeley, CA.

Grandcolas, Phillipe, and Cyrille D'Haese. (2001) The phylogeny of cockroach families: is the current molecular hypothesis robust? *Cladistics* 17: 48-55.

Greven, Harmut, and Nadine Zwanzig. (2013) Courtship, mating, and organization of the pronotum in the glowspot cockroach *Lucihormetica verrucosa* (Brunner von Wattenwyl,

Ellipsidion sp., Australian Capital Territory (CC-BY) Donald Hobern

Johnrehnia australiae, Australian Capital Territory (CC-BY) Donald Hobern

1865) (Blattodea: Blaberidae). *Entomologie heute* 25: 77-97.

Gurney, Ashley B. (1959a) New records of Orthoptera and Dermaptera from the United States. *Florida Entomologist* 42: 75-80.

Gurney, Ashley B. (1959b) The largest cockroach. *Proceedings Entomological Society Washington* 61(3): 133-134.

Gutiérrez, Esteban. (2004) Las cucarachas del género *Eurycotis* (Dictyoptera: Blattaria: Blattidae) en Puerto Rico, con la descripción de una especie nueva. *Solenodon* 4: 30-36.

Gutiérrez, Esteban, and Ariatna Linares. (2003) El género Cubano *Byrsotria* (Dictyoptera: Blattaria: Blaberidae: Blaberinae), con la descripción de una especie nueva. *Solenodon* 3: 31-48.

Helfer, Jacques R. (1963) *How to Know the Grasshoppers, Crickets, Cockroaches, and Their Allies*. Dover Publications.

Hintze-Podufal, C., and U. Nerling. (1986) Der Einfluss der nahrung auf entwicklung, wachstum und prareproduktionsphase von *Blaptica dubia* Stal. (Blaberiodea, Blaberidae). *Bulletin de la Societe Entomologique Suisse* 59: 177-186.

Holbrook, Glenn, and Coby Schal. (1998) Social influences on nymphal development in the cockroach, *Diploptera punctata*. *Physiological Entomology* 23: 121-130.

Hopkins, Heidi. (2014) A revision of the genus *Arenivaga* (Rehn) (Blattodea, Corydiidae), with descriptions of new species and key to the males of the genus. *Zookeys* 384: 1-256.

Hromadka, Jiri. (2005) *Brachynauphoeta brunneriana*. *Sklipkan Journal for Invertebrates Keepers and Breeders* 9(1): 33-34.

Hughes, J. H. (1949) *Aircraft and Public Health Service Foreign Quarantine Entomology*. Public Health Rep., supp. 210.

Husband, Robert W., and Barry M. O'Connor. (2003) A new genus and species of mite (Acari: Tarsonemina: Podapolipidae), ectoparasite of the Peruvian cockroaches, *Blaberus parabolicus* (walker) and *Eublaberus distanti* (Kirby) (Blattodea: Blaberidae). *International Journal of Acarology* 29(4): 331-338.

Hutchins, Ross E. (1966) *Insects*. Prentice Hall, Englewood Cliffs, NJ.

Inward, D., G. Beccaloni, and P. Eggleton. (2007) Death of an Order: A comprehensive molecular phylogenetic study confirms that termites are eusocial cockroaches. *Biological Letters* 3(3): 331-335.

Jefferson, G. T. (1958) A white-eyed mutant form of the American cockroach, *Periplaneta americana* (L.). *Nature* 182: 892.

Jeon, Alan. (2015) *Eurycotis lixa* Rehn, 1930. The black pearl of the Keys. *Invertebrates-Magazine* 14(4): 10-14.

John, B., and K. R. Lewis. (1958) Selection for interchange heterozygosity in an inbred culture of *Blaberus discoidalis* (Serville). *Genetics* 44(2): 253-267.

Kandilian, Kyle. (2011) 10 must-have roach genera for the hobbyist. *American Cockroach Society Special Publication 1*. pp. 25-30.

Kevan, D. K. McE. (1977) *The Higher Classification of the Orthopteroid Insects: A General View*. Lyman Ent. Mus. and Res. Lab. Mem. 4 (Special Pub. No. 12): 1-31.

Kevan, D. K. McE. (1980) Names involving the Madeira and Surinam cockroaches (Dictyoptera, Blattodea, Nauphoetidae) *The Entomological Record* 92: 77-82.

Kevan, D. K. McE. (1987) Distinguishing species of *Blaberus* without resource to disecction of genitalia. *Blattodea Culture Group Newsletter* 2: 1.

Kevan, D. K. McE., and L. Chopard. (1954) Blattodea from Northern Kenya and

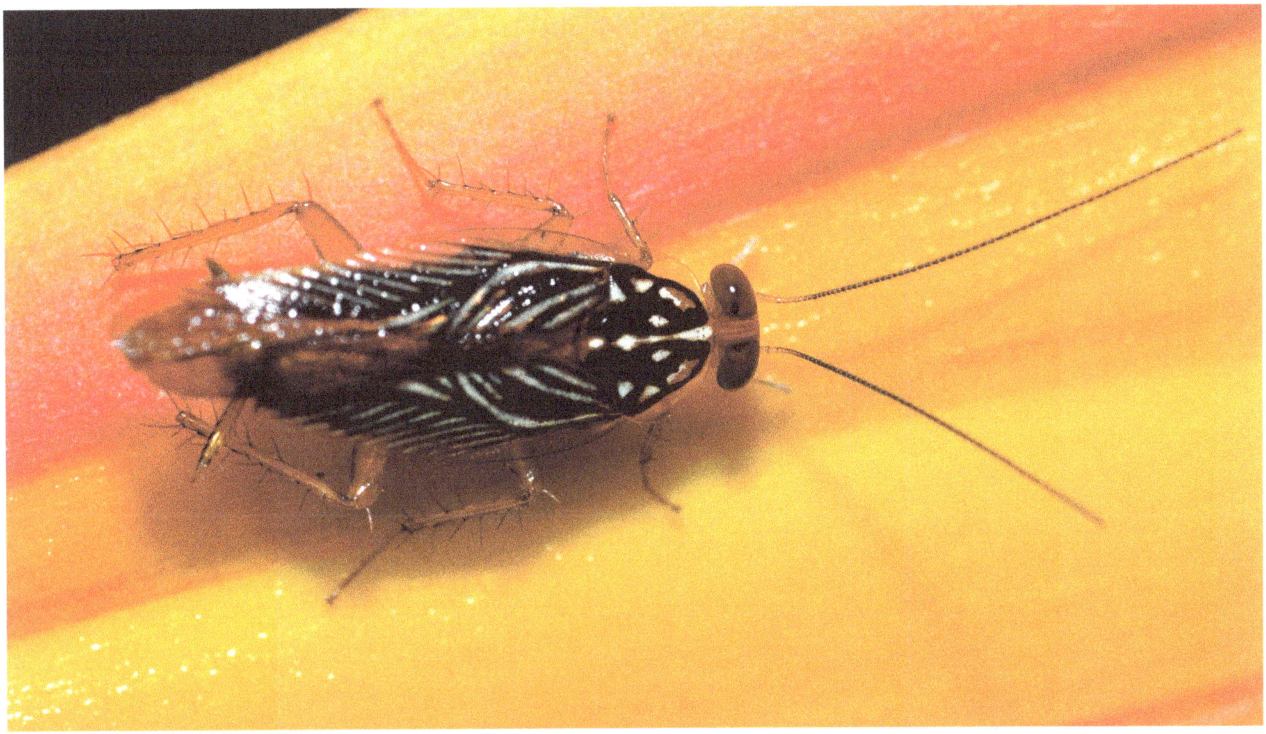

Ectobiidae, Costa Rica (CC-SA) Pavel Kirillov

Purple-eyed cockroach, *Lupparia* sp., Sulawesi, Indonesia (CC-SA) Bernard Dupont

Jubaland. *Ann. and Mag. Nat. Hist.*, Ser. 12. 7: 166-187.

Komatsu, Noriyuk. (2009) Allpet Roaches' roach care report of *Archiblatta hoeveni* Vollenhoven, 1862. *Invertebrates-Magazine* 9(3): 4.

Lynch, Susan V., et al. (2014) Effects of early-life exposure to allergens and bacteria on recurrent wheeze and atopy in urban children. *Journal of Allergy and Clinical Immunology* 134(3): 593-601.

Ludányi, Tibor. (2012a) Have your own alien queen in your insect room: keeping and breeding the emerald cockroach wasp. Part 1. *Invertebrates-Magazine* 11(3): 2-4.

Ludányi, Tibor. (2012b) Conquistador on six legs: keeping and breeding *Drymaplaneta semivittata*, the Gisborne roach. *Invertebrates-Magazine* 12(1): 2-4.

Mann, Darren J. (1988) *Blaptica dubia* Serville. *Blattodea Culture Group Newsletter* 3(2): 6.

Mann, Darren J. (1988) *Drymaplaneta semivitta* (Walker). *Blattodea Culture Group Newsletter* 3(2): 6.

Mann, Darren J. (1995) Species report number 10: *Pelmatosilpha larifuga* Gurney. *Blattodea Culture Group Newsletter* 9: 8-11.

Manning, Gail R. (1997) Rearing the pale-bordered field cockroach, *Pseudomops septentrionalis* Hebard (Blattellidae). *Invertebrates in Captivity Conference Proceedings*. pp. 60-61.

McKittrick, F. A. (1964) Evolutionary studies of cockroaches. *Cornell Univ. Agric. Exp. Stat., Memoir* 389: 1-197.

McMonigle, Orin. (2001) Featured invertebrate: *Gyna capucina*. *Invertebrates-Magazine* 1(1):6.

McMonigle, Orin. (2003) Madagascan hissing cockroaches. *Invertebrates-Magazine* 3(1): 6-10.

McMonigle, Orin. (2004a) Porcelain cockroach. *The Keeper* (International Fauna Society) (May): 1, 4.

McMonigle, Orin. (2004b) Featured invertebrate: Red and black cockroach *Opisthoplatia orientalis*. *Invertebrates-Magazine* 3(3): 14.

McMonigle, Orin. (2006a) The fattest cockroach. *Invertebrates-Magazine* 5(2): 9-11.

McMonigle, Orin. (2006b) Allpet Roaches' roach: *Dorylaea orini* Roth 2002. *Invertebrates-Magazine* 5(4): 16-17.

McMonigle, Orin. (2006c) Allpet Roaches' roach: Domino roach *Therea petiveriana*. *Invertebrates-Magazine* 6(1): 10-11.

McMonigle, Orin. (2007a) Allpet Roaches' roach: *Eurycotis decipiens* zebra cockroach. *Invertebrates-Magazine* 6(2): 10-11.

McMonigle, Orin. (2007b) Allpet Roaches' roach: *Deropeltis paulinoi* ornate velvet roach. *Invertebrates-Magazine* 6(3): 10-11.

McMonigle, Orin. (2007c) Allpet Roaches' roach: *Pseudomops septentrionalis* pale bordered field cockroach. *Invertebrates-Magazine* 6(4): 10-11.

McMonigle, Orin. (2007d) Allpet Roaches' roach: *Panchlora* sp. "giant." *Invertebrates-Magazine* 7(1): 14.

McMonigle, Orin. (2008a) Allpet Roaches' roach: *Blaberus giganteus* giant cave cockroach. *Invertebrates-Magazine* Vol. 7(2): 10-11.

McMonigle, Orin. (2008b) Allpet Roaches' roach: Death's head cockroach. *Invertebrates-Magazine* 7(3): 10-11.

McMonigle, Orin. (2008c) Allpet Roaches' roach: Rhinoceros cockroach *Macropanesthia rhinoceros*. *Invertebrates-Magazine* 7(4): 10-11.

McMonigle, Orin. (2008d) Allpet Roaches' roach: *Gromphadorhina oblongonota*, Van Herrewegue, 1973, wide horn hissing

Unidentified cockroach, Barro Colorado Island, Panama (CC-BY) Vince Smith

Sand cockroach (Polyphagidae), Israel (CC-BY) Eran Finkle

cockroach, ACS #39. *Invertebrates-Magazine* 8(1): 10.

McMonigle, Orin. (2009a) Allpet Roaches' roach: *Lucihormetica subcincta* (Walker 1868) glowspot cockroach. *Invertebrates-Magazine* 8(2): 14.

McMonigle, Orin. (2009b) Allpet Roaches' roach: *Panchlora nivea* (Linnaeus) 1758 green banana cockroach. *Invertebrates-Magazine* 8(3): 15.

McMonigle, Orin. (2009c) Allpet Roaches' roach: *Cryptocercus punctulatus*. *Invertebrates-Magazine* 8(4): 10.

McMonigle, Orin. (2009d) Allpet Roaches' roach: *Nauphoeta cinerea*, Olivier 1789 lobster cockroach. *Invertebrates-Magazine* 9(1): 10.

McMonigle, Orin. (2010a) Allpet Roaches' roach: *Diploptera punctata* (Eschscholtz, 1822) Pacific beetle mimic cockroach. *Invertebrates-Magazine* 9(2): 10.

McMonigle, Orin. (2010b) Allpet Roaches' roach: *Eurycotis floridana*. *Invertebrates-Magazine* 9(4): 10.

McMonigle, Orin. (2011a) Allpet Roaches' roach: *Aeluropoda insignis*, Butler 1882 flat-horn hissing cockroach. *Invertebrates-Magazine* 10(2): 9.

McMonigle, Orin. (2011b) Allpet Roaches' roach: *Macropanesthia rhinoceros* revisited. *Invertebrates-Magazine* 10(3): 13.

McMonigle, Orin. (2011c) Allpet Roaches' roach: *Dorylaea orini*. *Invertebrates-Magazine* 10(4): 17.

McMonigle, Orin. (2011d) Allpet Roaches' roach: *Oxyhaloa deusta* (1784, Thunberg) Cape redhead cockroach. *Invertebrates-Magazine* 11(1): 13.

McMonigle, Orin. (2011e) *Invertebrates for Exhibition Insects, Arachnids, and Other Invertebrates Suitable for Display in Classrooms, Museums, and Insect Zoos*. Coachwhip Publications, Landisville, PA.

McMonigle, Orin. (2012a) *Millipeds in Captivity: Diplopodan Husbandry and Reproductive Biology*. Coachwhip Publications, Landisville, PA.

McMonigle, Orin. (2012b) Allpet Roaches' roach: *Eurycotis opaca* var. "sunset" cockroach. *Invertebrates-Magazine* 11(3): 11-12.

McMonigle, Orin. (2012c) Allpet Roaches' roach: *Blaberus boliviensis* Princis, 1946 double dot cockroach. *Invertebrates-Magazine* 11(4): 11.

McMonigle, Orin. (2012d) Teneral mating as a standard function of copulation in the Dictyoptera. *Invertebrates-Magazine* 11(4): 2-4.

McMonigle, Orin. (2012e) Allpet Roaches' roach: *Eurycotis opaca* giant yellow margined. *Invertebrates-Magazine* 11(1): 12-13.

McMonigle, Orin. (2013a) Allpet Roaches' roach: Pallid cockroach *Phoetalia pallida*. *Invertebrates-Magazine* 12(1): 14-15.

McMonigle, Orin. (2013b) Allpet Roaches' roach: Pillbug cockroach *Corydidarum (Trichoblatta) pygmaea* *Invertebrates-Magazine* 12(2): 16-17.

McMonigle, Orin. (2013c) Allpet Roaches' roach: Simandoan cave roach *Simandoa conserfariam*. *Invertebrates-Magazine* 13(1): 16-17.

McMonigle, Orin. (2013d) *Breeding the World's Largest Arachnid Amblypygid Biology and Natural History*. Coachwhip Publications, Greenville, OH.

McMonigle, Orin. (2014a) Allpet Roaches' roach: Yellow spotted domino cockroach *Therea regularis*. *Invertebrates-Magazine* 13(2): 12-13.

McMonigle, Orin. (2014b) Allpet Roaches' roach: *Eublaberus* sp. "Pantanal" *Invertebrates-Magazine* 13(3): 12-13.

McMonigle, Orin. (2014c) Allpet Roaches' roach: *Paratemnopteryx couloniana*

Roach in Trinidad © Martin Gamache

Roach in French Guiana © Martin Gamache

Roach in French Guiana © Martin Gamache

Roach in French Guiana © Martin Gamache

(Saussure, 1863) ACS #100 little goblin. *Invertebrates-Magazine* 13(4): 12.

McMonigle, Orin. (2014d) Allpet Roaches' roach: *Ergaula pilosa* (Walker, 1868) ACS # 107 black beetle mimic. *Invertebrates-Magazine* 14(1): 13.

McMonigle, Orin. (2015a) Allpet Roaches' roach: *Paranauphoeta formosana* Matsumura, 1913. *Invertebrates-Magazine* 14(2): 12-13.

McMonigle, Orin. (2015b) Allpet Roaches' roach: Indonesian assassin mimic cockroach *Paranauphoeta discoidalis*. *Invertebrates-Magazine* 14(3): 12-13.

McMonigle, Orin. (2015c) Report of permanent *Androlaelaps schaeferi* exosymbiont transfer to *Blaberus craniifer* under laboratory conditions. *American Cockroach Society Publication* 2: 7-14.

McMonigle, Orin. (2015d) Cockroach eye color and documentation of 'lavender' eye color mutation in *Blaberus craniifer*. *American Cockroach Society Publication* 2: 25-26.

McMonigle, Orin, and Richard Willis. (2000, 2002) *Allpet Roaches: Care and Identification Handbook to the Pet and Feeder Cockroaches*. Elytra and Antenna, Brunswick, OH.

Mohan, Geoffrey. (2013) Military, not the Internet, blamed for invasive cockroach. *Los Angeles Times*.

Morgan, Randy. (1989) Culture corner: A house for termites. *Backyard Bugwatching* (6): 4-5, 14.

Morgan, Randy. (1999) Termite town: Homes for giant dampwood termites *Zootermopsis angusticollis* biology, husbandry, and display. *Invertebrates in Captivity Conference Proceedings*.

Olsen, Erik. (2015) *Therea olegrandjeani* in the terrarium and as feeders. *American Cockroach Society Publication* 2: 20-25.

Oonincx, D. G. A. B., and E. S. Dierenfeld. (2012). An investigation into the chemical composition of alternative invertebrate prey. *Zoo Biology* 31(1): 40-54.

Park, Andy. (1990) Guess who's coming to tea? *Australian Geographic Magazine* (18, Apr-Jun.)

Powell, J., and C. Hogue. (1979) *California Insects*. Univ. California Press, Berkeley, CA.

Preston-Mafham, Ken (1992) *Grasshoppers and Mantids of the World*. Sterling Publishing, New York.

Purpura, Dominick P. (1998) *Berta V. Scharrer 1906-1995 A Biographical Memoir*. National Academies Press, Washington D. C.

Rentz, David. (2014) *A Guide to the Cockroaches of Australia*. CSIRO Publishing.

Rose, Harley. (1989) *Selected Notes on the Giant Burrowing Cockroach*. University of Sydney. Agricultural Entomology.

Rose, Harley. (1990) *Care Instructions for the Giant Burrowing Cockroach*. The University of Sidney Department of Plant Pathology and Agricultural Entomology.

Rose, Harley, J. A. Walker, and J. R. Woodward. (2014) Five new species of soil burrowing cockroaches from Queensland (Blattodea: Blaberidae: Geoscapheinae). *Memoirs of the Queensland Museum* 59: 11-23.

Rosenstreich, David L., et al. (1997) The role of cockroach allergy and exposure to cockroach allergen in causing morbidity among inner-city children with asthma. *The New England Journal of Medicine* 336: 1356-1363.

Ross, M. H. (1993) Hybridization studies on *Blattella germanica* and *B. asahinai* (Dictyoptera: Blattellidae): Species divergence and a possible influence of a major chromosome mutation. *Genome* 36(2): 278-88.

Roth, Louis M. (1967) The evolutionary significance of rotation of the ootheca in the Blattaria. *Psyche* 74: 85-103.

Ellipsidion australe nymph, Australia (CC-BY) Graham Wise

Ellipsidion australe nymph, Australia (CC-BY) James Niland

Roth, Louis M. (1968) Oothecae of the Blattaria. *Annals of the Entomological Society of America* 61(1): 83-111.

Roth, Louis M. (1970) The male genitalia of Blattaria IV. Blaberidae: Blaberinae. *Psyche* 77: 308-336.

Roth, Louis M. (1977; 1979; 1979; 1982) A taxonomic revision of the Panesthiinae of the world, parts I to IV. *Australian Journal Zoology,* Suppl. Serial. Nos. 48; 69; 74; 82.

Roth, Louis M. (1981) The mother-offspring relationship of some Blaberid cockroaches (Dictyoptera: Blattaria: Blaberidae). *Proc. Entomol. Soc. Wash.*, 83(3): 390-398.

Roth, Louis M. (1981) in Introduction to *The American Cockroach* by Bell, W. J., and K. G. Adiyodi. Springer Science .

Roth, Louis M. (1991) A new cave-dwelling cockroach from Western Australia (Blattaria: Nocticolidae). *Records of the Western Australian Museum* 15(1): 17-21.

Roth, Louis M. (1995) *Africalolampra ehrmanni* new genus and species, and the male of *Paraplecta parva* Princis (Blattaria: Blaberidae). *Psyche* 102(1-2): 89-98.

Roth, Louis M. (1997) *Pseudobalta*, a new Australian ovoviviparous cockroach genus (Dictyoptera: Blattaria: Blattellidae: Pseudophyllodromiinae). *Australian Journal of Entomology* 36: 101-108.

Roth, Louis M. (1998) The cockroach genus *Pycnoscelus* Scudder, with a description of *Pycnoscelus femapterus*, sp. nov. (Blattaria: Blaberidae: Pycnoscelinae). *Oriental Insects* 32: 93-130.

Roth, Louis M., and Piotr Naskrecki. (2004) A new genus and species of cave cockroach (Blaberidae: Oxyhaloinae) from Guinea, West Africa. *Jour. of Orthoptera Res.* 13(1): 57-61.

Roth, Louis M., and Edwin R. Willis. (1954) The reproduction of cockroaches. *Smithsonian Miscellaneous Collections* 122: 12.

Roth, Louis M., and Edwin R. Willis. (1955a) The water content of cockroach eggs during embryogenesis in relation to oviposition behavior. *Journal of Experimental Zoology* 128: 489-509.

Roth, Louis M., and Edwin R. Willis. (1955b) Intra-uterine nutrition of the "beetle-roach" *Diploptera dytiscoides* (Serv.) during embryogenesis, with notes on its biology in the laboratory (Blattaria: Diplopteridae). *Psyche* 62(2): 55-68.

Roth, Louis M., and Edwin R. Willis. (1956) Parthenogenesis in cockroaches. *Annals of the Entomological Society of America* 49: 195-204.

Roth, Louis M., and Edwin R. Willis. (1957a) The medical and veterinary importance of Cockroaches. *Smithsonian Miscellaneous Collections* 134: 1-147.

Roth, Louis M., and Edwin R. Willis. (1957b) The biology of *Panchlora nivea*, with observations on the eggs of other Blattaria. *Transactions of the A.E.S.* 83(4): 195-207.

Roth, Louis M., and Edwin R. Willis. (1960) The biotic associations of cockroaches. *Smithsonian Miscellaneous Collections* 141: 1-470.

Rugg, Douglas, and Harley Rose. (1991) Biology of *Macropanesthia rhinoceros* Saussure (Dictyoptera: Blaberidae). *Annals of the Entomological Society of America* 84(6): 575-582.

Sastre, J., et al. (1996) Allergy to cockroaches in patients with asthma and rhinitis in an urban area (Madrid). *Allergy* 51: 582-586.

Saupe, R. (1928) Zur kenntnis der lebensweise der riesenschabe *Blabera fusca* Brunner und der gewächshausschabe *Pycnoscelus surinamensis* L. Zeitschr. *Angew. Ent.* 14: 461-500.

Schaefer, C. W., and D. B. Peckham. (1968). Host Preference studies on a mite infesting the cockroach *Gromphadorhina portentosa*.

Annals of the Entomological Society of America 61: 1475-1478.

Schal, Coby, Jean Fraser, and William J. Bell. (1982) Disturbance stridulation and chemical defence in nymphs of the tropical cockroach *Megaloblatta blaberoides*. *Journal of Insect Physiology* 28: 541-552.

Schal, Coby, J. Y. Gautier, and William J. Bell. (1984) Behavioral ecology of cockroaches. *Biological Review* 59: 209-254.

Schweid, Richard. (1999) *The Cockroach Papers: A Compendium of History and Lore*. Basic Books, New York.

Shelford, Robert. (1907) Aquatic roaches. *Zoologist* 4: 221-226.

Stanek, V. J. (1969) *The Pictorial Encyclopedia of Insects*. Hamlyn House, London.

Tanaka, Akira, and Tamiyo Ito. (1997) Studies of the genetics and expression of Prowing (Pw): A primitive homeotic mutant of the German cockroach, *Blattella germanica*. *Zoological Science* 14: 339-346.

Takahashi, R. (1926) Observations on the aquatic cockroach *Opisthoplatia maculata*. *Dobuts Zasshi*. (Tokyo) 38: 89-92.

Thorne, Barbara L., Nancy L. Breisch, and Michael I. Haverty. (2002) Longevity of kings and queens and first time of production of fertile progeny in dampwood termite (Isoptera; Termopsidae; Zootermopsis) colonies with different reproductive structures. *Journal of Animal Ecology* 71(6): 1030-1041.

Thomson, Arthur J. (1921) Black-beetles. *The Living Age* 310(4029): 223.

Turner, Clive. (1995) Collecting cockroaches in Greece. *Blattodea Culture Group Newsletter* 10: 2-3.

Van Herrewege, C. (1973) Contribution a l'etude des Blattaria de la Faune Malgache. I. Princisia, gen. nov., voisin de *Gromphadorhina* Brunner v. w.: Description d'une Espece Nouvelle. *Bulletin Mensuel Soc. Linnéenne de Lyon* 10: 262-267.

Van Herrewege, C. (1980) Contribution a l'etude des Blattaria de la Faune Malgache. II Describtion de huit Especes Nouvelles Appartenant aux Genres *Gromphadorhina* Brunner v. W. et. *Elliptorhina* gen. nov. *Bulletin Mensuel Soc. Linnéenne de Lyon* 42: 150.

Vishniakova, V. N. (1968) Mesozoic Blattids with external ovipositors and details of their reproduction. In: *Jurassic Insects of Karatau*; Ed., B.B. Rohdendorf. Akademiya Naukk SSSR, *Ordelenie Obschei Biolgii* pp. 55-86.

Vršanský, Peter. (1997) *Piniblattella* gen. nov., the most ancient genus of the family Blattellid (Blattodea) from the Lower Cretaceous of Siberia. *Entomological Problems* 28: 67-79.

Vršanský, Peter, et al. (2012) Light-mimicking cockroaches indicate Tertiary origin of recent terrestrial luminescence. *Naturwissenschaften* 99(9): 739-749.

Wang, Zong-Qing, et al. (2015) The wood-feeding genus Cryptocercus (Blattodea: Cryptocercidae), with description of two new species based on female genitalia. *Florida Entomologist* 98(1): 260-271.

Wedberg, S. E., C. D. Brandt, and C. F. Helmboltd. (1949) The passage of microorganisms through the digestive tract of *Blaberus craniifer* mounted under controlled conditions. *Journal of Bacteriology* 58: 573-578.

Wheeler, W. M. (1939) The dry rot in academic biology. Chapter VI *in* 'Essays in Philosophical Biology.' Harvard University Press: Cambridge, MA.

Willis, Edwin R., George R. Riser, and Louis M. Roth. (1958) Observations on reproduction and development in cockroaches. *Annals of the Entomological Society of America* 51: 53-69.

Eublaberus posticus © Alex Yelich

Eublaberus posticus © Alex Yelich

Cave cockroaches (*Pycnoscelus indicus*) in bat guano, Sarawak, Malaysia (CC-SA) Bernard Dupont

Nyctibora nymph, Barro Colorado Island, Panama (CC) Katja Schulz

"Sun Tiger," *Platyzosteria* species nymph © Bruce Thomson

Wyttenbach, Robert, and Thomas Eisner. (2001) Use of defensive glands during mating in a cockroach (*Diploptera punctata*). *Chemoecology* 1: 25-28.

Zha, Chen. (2015) Captive breeding of *Eucorydia aenea dasytoides* (Walker 1868), the Taiwanese flower cockroach. *American Cockroach Society Publication* 2: 4-7.

Zompro, Oliver, and Ingo Fritzsche. (1999) *Lucihormetica fenestrata* n. gen. n. sp. The first record of luminescence in an Orthopteroid insect (Dictyoptera: Blaberidae: Blaberinae: Brachycolini). *Amazoniana* XV (3/4).

Gyna caffrorum, Kruger NP, South Africa (CC-SA) Bernard Dupont

Catara rugosicollis, Sarawak, Malaysia (CC-SA) Bernard Dupont

Colorful cockroach, Heredia, Costa Rica (CC-SA) Llisa

Balta notulata, Christmas Island, Australia (CC) John Tann

Also Available by Orin McMonigle

CENTIPEDES IN CAPTIVITY

MILLIPEDS IN CAPTIVITY

ASSASSIN BUGS,
WATERSCORPIONS,
AND OTHER
HEMIPTERA

PILLBUGS AND
OTHER ISOPODS

Also Available by
Orin McMonigle

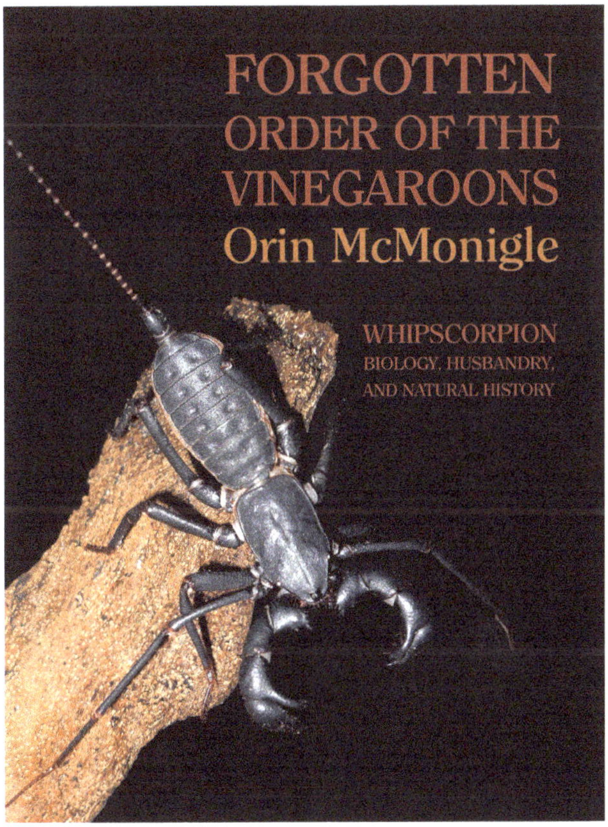

FORGOTTEN ORDER OF THE VINEGAROON

BREEDING THE WORLD'S LARGEST LIVING ARACHNID

www.ingramcontent.com/pod-product-compliance
Ingram Content Group UK Ltd.
Pitfield, Milton Keynes, MK11 3LW, UK
UKHW060225061026
15625UKWH00008B/498

9 781616 464271